X

comedy, lighthearted adventure, travel, and natural history that is uniquely Gerald Durrell's.

The author of eighteen previous books, Gerald Durrell is one of the United States' and Britain's most popular writers on travel and natural history. He was born in Jamshedpur, India, in 1925. His interest in zoology began when his family moved to the island of Corfu in the 1930s. For the past forty years, Durrell has organized and led animal-collecting expeditions all over the world. He is the founder and Honorary Director of the Jersey Wildlife Preservation Trust.

How to Shoot an Amateur Naturalist

By the same author

THE OVERLOADED ARK
THREE SINGLES TO ADVENTURE
THE BAFUT BEAGLES
THE DRUNKEN FOREST
MY FAMILY AND OTHER ANIMALS
ENCOUNTERS WITH ANIMALS
A ZOO IN MY LUGGAGE
THE WHISPERING LAND
TWO IN THE BUSH
BIRDS, BEASTS AND RELATIVES
FILLETS OF PLAICE
CATCH ME A COLOBUS
BEASTS IN MY BELFRY
THE STATIONARY ARK
GOLDEN BATS AND PINK PIGEONS
THE GARDEN OF THE GODS
MENAGERIE MANOR
THE PICNIC AND SUCHLIKE PANDEMONIUM

Novels

ROSY IS MY RELATIVE
THE MOCKERY BIRD

For children

THE NEW NOAH
ISLAND ZOOS
LOOK AT ZOOS
MY FAVOURITE ANIMAL STORIES
THE DONKEY RUSTLERS
THE TALKING PARCEL

Gerald
Durrell

HOW TO SHOOT AN AMATEUR NATURALIST

LITTLE, BROWN AND COMPANY
BOSTON TORONTO

FIRST AMERICAN EDITION

Library of Congress Cataloging-in-Publication Data

Durrell, Gerald Malcolm, 1925–
How to shoot an amateur naturalist.

1. Natural history. I. Title.
QH45.5.D87 1985 508 85-10197
ISBN 0-316-19717-3

MV

PRINTED IN THE UNITED STATES OF AMERICA

This book is for
Paula,
Jonathan
and
Alastair
with love and respect

A Word in Advance

It is perhaps necessary that I give some rational explanation of the somewhat bizarre title for this book.

If you look up the word 'shoot' in the dictionary, you will see that among its many meanings, ranging from scoring a goal to the growth of a plant, is the phrase 'to photograph, especially for motion pictures'. Thus did this title take shape, for this is a chronicle of the year-long shooting my wife, Lee, and I undertook to complete thirteen half-hour television programmes called *The Amateur Naturalist*.

Some time ago, I was approached with the idea of writing a book to be called (it was suggested) 'The Complete Amateur Naturalist'. I immediately objected to the word 'complete'. I said that anyone writing a guide claiming to be complete was asking for trouble, and that to use that word in connection with the world of nature – where we are making discoveries with such rapidity that we scarcely have time to record them – would be dangerous, to say the least. So it was decided to call it simply 'The Amateur Naturalist'.

To begin with, this book was conceived as a small guide confining itself to the British Isles. Then somebody said, wouldn't it be fun to include Europe; then someone else spoke eloquently of the need for such a book in America, and still another of the awful need for such a book in Australia and New Zealand, South Africa and points north, south, east and west. The whole thing then got out of hand.

I knew I could not write such a book *and* do the research for it, so I suggested to Lee that she should stop being merely decorative and take her PhD (unused since our marriage) out for an airing and superintend the research on the project, which now threatened to exceed in size and scope the *Encyclopaedia Britannica*. This she dutifully did, and as well as working out the shape of the book (we had to divide it into ecosystems instead of unwieldy and unbiological countries) she started on the mammoth task of combing a thousand books, checking and rechecking (you have no idea how scientists contradict one another) and phoning up a plethora of pundits, seeking advice.

As this stream of information landed on my desk, my job was to turn it into what Lee, rather insubordinately, calls my 'purple prose'.

The book took a little over two years to complete, and that it did not end in divorce says much for Lee's patience and forbearance. It was an immediate success, and we sat back glowing self-righteously and thought we had earned a holiday. However, everyone was so delighted with the book that before we knew it we had agreed to do a television series based on it; and this, nearly eighteen months later, we have just completed.

Our producer for the series was to be Paula Quigley, known to us as Quiggers, whom we knew and loved and had worked with in Mauritius and Madagascar while making the television series *Ark on the Move*. She is slender, petite, with a mop of dark curly hair, a snub nose like a Pekinese and those curious eyes that can be both blue and green, depending on what she is wearing. She is also possessed of unfairly long eyelashes that can only be equalled by a giraffe. In addition to her normal, very pleasantly feminine soprano speaking voice, she is capable of a bellow that would have won her first prize in any town-crier contest, and it came in extremely useful as we had not budgeted for walkie-talkie or a megaphone. (With Paula as part of the team we did not need either of these adjuncts of communication.)

We were to have two directors, Jonathan Harris and Alastair Brown. Alastair was to direct seven of the programmes and Jonathan six. Alastair looked very donnish, since his receding hairline gave his brow noble proportions. Behind his glasses his pale-blue eyes held a mystical gleam that you would have expected from the White Knight, and he wore an almost perpetual grin. He had a habit of holding his head on one side and revolving slowly in a circle, so that he bore a vivid resemblance to the Hanged Man on a Tarot card. His habit of speaking in half-sentences which did not appear to have any relation to each other made communication difficult, but fortunately we had Paula to act as translator on those occasions when Alastair got so over-excited that he appeared to be speaking Patagonian. As a contrast, Jonathan was dark and somewhat glowering, handsome in a Heathcliffian sort of way, with a husky voice and a meticulous way of speaking that at first made you think him pedantic, until you discovered his puckish sense of humour.

Whether or not it was a good idea to have two directors is open to doubt. True, it leads to a jolly spirit of competition: but directors, when you give them their head, tend to get overenthusiastic, and in this particular instance they both vied with each other to see which one could get us to perform the most dangerous and hair-raising feats, and if it had not been for the tender care that Paula took of us we might both have disappeared from life, for once a director gets a fixation about a scene nothing will deter him and you soon find out that you are considered expendable. This attitude is summed up by Alfred Hitchcock's remark, 'I didn't say actors and actresses were cattle, I said they should be treated like cattle.' At any rate, now is my chance to get my own back.

Added to the hazards of the whole venture was the fact that Paula, Alastair and Jonathan were not naturalists. As we soon discovered, their knowledge of the natural world around them would have fitted into an egg-cup and left room for an egg. They could, with much difficulty and after considerable thought, just distinguish a mouse from a giraffe, a crab from a shark, a frog from a boa constrictor and a butterfly from an eagle, but it was an uphill struggle. However, as the series progressed, we found that they were actually becoming converted and turning into naturalists, however rudimentary, and this gave us hope for the series, as it was designed to get everyone, everywhere, from nine to ninety, to open their eyes and become amateur naturalists.

We are glad we did the series, in spite of the many difficulties that we all encountered. Whether we would have attempted it if we had realised what it would entail is very doubtful. However, to visit different parts of the world at other people's expense is one of the delights of life, and it was especially so for me because, although I had previously seen a number of the things that we saw and filmed, Lee had not and so it was a pleasure to watch her pleasure.

To film these thirteen shows in twelve months we travelled 49,000 miles from the Canadian Rockies to Panama, from South Africa to the northernmost tip of the British Isles. Finally, for those people who think that we lead an exotic existence, let me say that, while it is fascinating to travel like this, the making of thirteen half-hour television programmes is damned hard work and very exhausting. If you can all remain friends at the end of it, it is one of life's major miracles.

Terms

SPRINKLED throughout this book are a few film terms. At the risk of boring the reader, I have defined here some of the more common ones so that I do not have to make the text unnecessarily laborious by explaining them.

1. *The talent.* Lee and I, or anyone daft enough to be a presenter of a series.

2. *Close-up.* Face filling the screen showing the wear and tear of years of dissipation. Showing Lee looking beautiful.

3. *Medium shot.* A picture of you roughly from the knees up, depicting clearly the ravages of delicious meals and fragrant wines over the years. Lee looking slim, like a very smug minnow.

4. *Long shot.* Mercifully, both of us almost out of sight. Lots of trees, mountains and other bits of nature covering our defects.
5. *Pan shot.* The camera follows you relentlessly from right to left or vice versa, while you walk (or stumble) trying to remember your lines and not get your microphone cord tied up in the undergrowth.
6. *Zoom shot.* You are half a mile away from the camera and, when you least suspect it, the camera appears to come within about two feet of your face, revealing that you haven't combed your beard. You also, by virtue of the lens used, look infinitely more debauched than you thought you were. Lee, infuriatingly, looks better than Jackie Onassis after a long rest, a steam bath and a luxurious massage.
7. *Voice or piece to camera.* You try to look at the camera as if it was your favourite friend and speak to it lovingly, without forgetting your lines.
8. *Fluff.* When you are doing a voice to camera and you say 'I want you particularly to notice owl this how . . .' instead of 'I want you particularly to notice how this owl . . .' or any similar gaffe. Unfortunately, fluffs have a habit of breeding and, when you have cured one, another one develops, until you find you are talking gibberish and have to go and lie down under a tree and be cosseted by your wife and hated by your director.
9. *Budget.* A sum of money carefully designed to be too small for the task in hand.
10. *Hair in gate.* When you have just (at the fifteenth attempt) successfully completed a highly complicated scene, the cameraman tells you that a large portion of the surrounding countryside has mysteriously found

its way into the 'gate' of the camera, ruining the film and necessitating a repeat performance, during which you 'fluff' and forget your lines. On occasion camera-men have been known to have their throats slit by infuriated talent and directors for this fault.

SHOOT ONE

Nîmes, in Provence, was suffering a heat wave. People wilted, the hot air in their lungs seeming useless to sustain life. The city, with its wide, tree-lined boulevards and its network of narrow alleyways redolent with fresh bread, drains, fruit, vegetables and cats, lay under the sun, shimmering and cooking gently.

In its centre lay the great Roman arena, like some medieval crown, rescued after a millennium in the sea, carunculated and blurred by the corals. It glittered in the fierce glare of the sun and, in every nook and cranny of this crown, pigeons gasped and gaped in the shade. Dogs panted from tree to tree, fronds of white saliva trailing from their tongues. Cicadas relentlessly fretted on the great patched bodies of the plane trees that lined the boulevards. Ice in the glasses at the cafés melted as you watched. How hot was it? Hot enough to roast an ox on the sands of the arena. Hot enough to poach an egg in the pools and ponds of the Jardin de la

Fontaine. Warm enough to make toast on the tiles of every building in Nîmes – or so it appeared to us. It was 100° and over in the shade, and your body felt like damp rubber, smelly and unliveable-in. It was Sidney Smith who once complained that it was so hot he had to take off his flesh and sit down in his bones. You knew how he felt.

On the outskirts of the city, where the tinder-dry Garrigue begins, we were giving our house a face-lift. Walls were knocked down, tiles relaid, doors refitted. In this heat, it was difficult to argue and expostulate with masons, carpenters and plumbers, all as heat-stunned as you were, drunk and exhausted by the sheer effort of dragging hot air in and out of your lungs. It was at that moment that we heard that all the contracts had been signed for the making of *The Amateur Naturalist* and shooting was to begin immediately. The first programme was to deal with animal life that inhabited cliffs and rocky shores, and in it we wanted to show that the sea's edge and a cliff-face are both divided into zones in which different creatures live.

They say contrasts are good for you. Well, we had a contrast. We left the shimmering city of Nîmes, we left the twanging Punch-and-Judy accents of Provence and we flew up to the most northerly point of the British Isles, to the island of Unst in Shetland, where the air was bland and only as warm as fresh milk, and the accents were as blurred as the gentle noises of humble-bees.

After the usual indescribable mess which is inevitable when you entrust your arrangements to those experts, the travel agents, we suddenly found ourselves flighting over a landscape of muted pastel green, and then landing, several degrees cooler, on the tarmac at Aberdeen. Here we met up with the crew. Chris, the cameraman, was short, stocky and bearded, with an air of complete competence, looking like one of the more endearing illustrations of gnomes in the wonderful Gnome Book. Brian, our sound-recordist, looked with his curly dark hair, his well-groomed appearance, more as if he were a bank manger in, say, Penge or Surbiton than somebody who was willing to lie in the bushes magically trapping the sounds of life. Chris's assistant, a young good-looking lad called David, gave me somewhat of a shock. As he approached us through the airport lounge he appeared to be in the last stages of a nervous disease akin to, but more severe than, St Vitus' dance. With this affliction it seemed to me curious to make the boy an assistant cameraman. It was only when he got closer and I discovered that he was dancing to some tribal music on his Sony Walkman that my sympathy for his affliction abruptly ended.

Aberdeen is a lovely, neat city with its solemn-faced houses, wearing roofs of grey slates like Beatle-styled hair-cuts, streets lined with great beds of roses with huge multi-coloured petals, silk-soft, feasting the eye and the nose. I was delighted that, because of the complications of getting to Shetland on the right date, we were forced to fly from Aberdeen to Lerwick, the southernmost tip of the island, and then make our way in a minibus by road to Unst, crossing by two ferries *en route* as a bonus.

It was the colouring that first struck you. The gentleness of the colours, was as though each green or brown had been muted and softened by an appliqué of chalk, and the clouds, low and sculpted to the exact shade of grey and very pale coffee of the tangles of sheep's wool that hung on the fences and in spiky thickets. The rolling, low hills were pale, creamy emerald or, where the heather grew, a rich chocolatey-mauve. The hedges along the way were golden with buttercups and dandelions, and in places purple loose-strife blazed and in the damp hollows golden iris bloomed like banners in an army of green sword-shaped leaves. For some reason, it reminded me of New Zealand with its rolling empty landscape and roads with practically no traffic and the same sense of remoteness. In places the heather was sabre-slashed where the peat had been cut out in pieces from the land. These bricks of peat, rich and dark as plum cake, lay drying in great jumbled piles beside the tiny crofts. We came at last to our motel on the shores of the sea and here, once we were installed in our room, Jonathan joined us, thoughtfully bringing with him a bottle of pale-yellow Glenmorangie, nectar of the Gods.

'Now,' said Jonathan, after he had sipped approvingly, 'tomorrow we go up to the white rocks – that's the headland of Hermaness and a great gannet colony. So we'll climb down the cliff—'

'Just a minute,' I interrupted. 'What cliff? Nobody said anything about a cliff to me.'

'It's just a cliff,' said Jonathan airily. 'All the different species of bird breed on it – guillemots, puffins, kittiwakes and so on. It's one of the biggest breeding colonies of seabirds in the Northern Hemisphere.'

'What about this cliff?' I asked, not to be distracted.

'Well, we've got to get *down* it,' said Jonathan, 'or we can't film the birds.'

'How high is it?'

'Not really high,' he replied evasively.

'How high?'

'Oh, about . . . four or five hundred feet,' he said; and then, seeing my expression, he added hastily: 'There's a perfectly good path down; wardens use it all the time.'

'I have told you that I suffer from vertigo, Mr Harris, have I not?' I asked.

'You have.'

'I know that it is a very stupid thing to suffer from, but I can't help that. I have tried to cure myself, but I can't do it. If I have my shoes resoled, I am dizzy for about a fortnight. That's how bad it is.'

'I sympathize,' said Jonathan untruthfully, 'but you'll be OK. It's as easy as falling off a log.'

'I can't congratulate you on your choice of metaphor,' I said acidly.

The next morning, to everyone's astonishment, the clouds had been swept away during the night and we were under a sky that was almost Mediterranean blue in its brilliance. Jonathan was ebullient.

'Wonderful day for filming,' he said, staring at me owlishly through his spectacles. 'How are you feeling?'

'If, by that enquiry, you mean has my vertigo miraculously disappeared during the night, the answer is no.'

'You'll be all right,' he said uneasily. 'Honestly, the paths are perfectly OK. They use them all the time, and no one has ever had an accident.'

'I would hate to create a precedent,' I said.

We drove as far as we could, and then started over the slopes of heather and emerald-green grass towards the great cliffs of Hermaness. In amongst the heather, sundews raised innocent sticky faces to any passing insect, ready to trap and engulf it, so many thumbnail octopi growing among the twisted witches'-broom heather roots. Over the green grass, cropped like a bowling-green by the grazing sheep, cotton-grass grew in great profusion. From a distance, this looked like fields of snow, but when you got close and walked through it and it was blown by the wind it was like a million rabbit scuts, flicking and glinting.

Overhead, the chocolate-dark skuas wheeled on huge wings, keeping a careful eye on us, for they had young concealed in the heather. We discovered one youngster, the size of a small chicken. Clad in pale tawny down, with his black face and beak and huge, dark, soulful eyes, he was an enchanting baby. Lee and I pursued him as he waddled off through the heather and his parents started to dive-bomb us – a really impressive performance. Huge wings taut, they swooped at us, the wind whooshing through their feathers, looking like strange, coffee-coloured Concordes. At the very last minute, within a few feet of your head, they would veer away, fly round and come in again. Lee had by now caught the baby, so the parents concentrated their attacks on her. As I knew that skuas were capable of knocking a man down with a clout from one of their wings, I relieved her of the baby and the parents turned their attention to me, getting closer and closer with each swoop, the wind purring through their wings as they dived. At first, I instinctively ducked each time, but then I discovered that if you let them get to within a dozen feet or so, and then waved your arms at them, they would sheer off.

'Let us,' said Jonathan, 'do a piece to camera about skuas, with that enchanting baby sitting in your lap.'

So the camera was set up and a microphone concealed around my neck. All this activity made the parent skuas twice as distraught as they had been and they redoubled their attacks, dive-bombing now me and then the camera, getting dangerously close. When the camera was ready, I squatted down in the heather and placed the fat baby in my lap. I had just opened my mouth to start on a fascinating lecture about skuas, when the baby stood up suddenly, pecked my thumb unexpectedly, making me lose the thread of my discourse, and then proceeded to defecate loudly and copiously all over my knees.

'Nature white in tooth and claw,' said Jonathan, as I mopped the glutinous, fishy mess off my trousers with my handkerchief. 'I don't think we can use *that* shot in the film.'

'When you have finished laughing,' I said to Lee, 'you might like to take this damn baby and release it. I think I've been intimate enough with skuas for one day.'

She took my fat, fluffy friend and placed him in the heather some twenty feet away. He took off at a spread-legged, crouching, flat-footed run looking remarkably like an elderly fat lady in a fur coat, pursuing a bus.

‘He is cute,’ said Lee wistfully. ‘I wish we could have kept him.’

‘I don’t,’ I said. ‘We wouldn’t have been able to afford the dry-cleaning bills.’

Skuas, of course, are one of the most graceful predators of the sky. Like sun-bronzed pirates, they pursue other birds, harrying them ceaselessly until they are forced to disgorge the fish they have caught. Then the skua swoops and snatches the treasure in mid-air. They are such bold buccaneers that they have even been known to grab a gannet’s wing-tip in order to get it to relinquish a fish. Skuas will eat anything and are not at all averse to stealing fish from a parent bird, be it gannet or guillemot, and then feasting on their eggs and young as well.

We moved on, the flocks of sheep like clotted cream on the green baize of the turf, the sun brilliant above us. We had come muffled up against the reputedly inclement weather of the Shetland Islands and now found ourselves sweating and discarding coats and pullovers. Presently, the land started to drop away to precipitous cliffs and beyond was the Atlantic, blue as gentian flowers. Wheatears were everywhere, their rumps flashing like little white lights as they danced ahead of us. Two ravens, black as mourning-bands, flew slowly along the edge of the cliffs, cronking at one another dolefully. High in the sky, a lark hung and poured forth its wonderful liquid song. If a shooting star could sing, I believe it would sing like a lark.

Soon we came to the cliff-edge. Some six hundred feet below us, the great smooth blue waves shouldered their way in between the rocks in a riot of spray like beds of white chrysanthemums. The air was full of the surge of surf and the cries of thousands upon thousands of seabirds that drifted like a snowstorm along the cliffs. The mind boggled at the numbers. Hundreds upon hundreds of gannets, kittiwakes, fulmars, shags, razorbills, gulls, skuas, and tens of thousands of puffins Could the sea possibly hold enough fish to feed this cacophonous aerial army and its numerous families that lined the cliff-faces?

At the cliff-edge, where the earth was soft enough for digging, was the puffins' special area. Here they excavated their burrows with powerful beaks and feet. They sat around in their hundreds, almost letting you tread on them before launching themselves over the cliff's edge and flying away with rapid wing-beats, their feet trailing behind them like little orange ping-pong bats. To see the green cliff-tops lined with hundreds of these comical waddling birds, each very upright in its neat black and white dinner-jacket, each wearing (as it were) one of those carnival noses, a huge beak striped orange and red. It was like watching a convention of clowns. Many of them, to add to their ridiculous aspect, had just flown in from fishing far out at sea (for they travel as far as three hundred kilometres away to fish) bearing in their brightly painted beaks handfuls of sand-eels. These were carefully arranged across the beak, hanging down each side like a fishy moustache. The extraordinary thing was that the sand-eels were arranged head to tail like sardines in a tin. How the birds manage to catch the eels and arrange them in this meticulous way is an extraordinary feat.

Further along the cliff, we came upon two men engaged in an extremely curious pursuit – puffin fishing. I know that in remote corners of the world the inhabitants sometimes become very eccentric, but I have never seen anything to equal this. Seated on their behinds, they shuffled slowly down the turf towards the cliff's edge where the solemn-faced puffins congregated, eyeing the men's approach warily. The first man carried a long pole, on the end of which was a noose. Having chosen his puffin, he then cautiously slid towards it, manoeuvred the noose carefully until it got round one of the bird's orange feet, then pulled the flapping, squawking bird towards him; and, when it was within range, the other man seized it. I thought this a curious way to treat birds in what was, after all, a sanctuary. However, as we got closer, I could see that they were fastening a ring to the puffin's leg. These rings are the puffin's passport or identity card. If it is picked up sick or dead or merely caught in a net, the ring tells you where it came from and at what date. It is a sort of bureaucracy for birds, but it does add to our knowledge of the mysterious lives seabirds lead, far from shore, in the non-breeding season.

The two wardens told us that there were a hundred thousand puffins breeding on the cliff of Hermaness and it was only during the breeding season when the birds were ashore that they could catch them for ringing. They handed me their captive so that I could do a piece to camera on the mysteries of puffins and I discovered very quickly that puffins may look comical and a bit dim-witted but they certainly know how to defend themselves. I caught hold of him rather carelessly, and in a moment the thick, razor-sharp beak had snapped shut on my thumb like a great rat-trap and my hands were being torn to shreds by the bird's needle-sharp claws, as sharp as any cat's. After I had done my piece to camera, I was only too pleased to release my belligerent co-star and let Lee do some first aid on my lacerated hands.

'Now that I have been almost disembowelled by a puffin,' I said to Jonathan, 'what other treats do you have in store for me?'

'Now we go down the cliff,' said Jonathan.

'Where?' I asked.

'Here,' he said, pointing to the cliff-edge that, as far as I could see, dropped sheer, six hundred feet to the sea below.

'But you said there was a path,' I protested.

'There is,' he said. 'If you go to the edge, you'll see it.'

Gingerly, my stomach turning over, I approached the edge. Meandering down among the tussocks of grass and thrift was a faint line that looked as though once, in the dim and distant past, a flock of inebriated goats had staggered down the cliff-face to indulge in God knows what alcoholic orgy.

'Call that a path?' I enquired. 'If I were a chamois, I might agree with you, but no man born of woman could go down that.'

As I spoke, Chris, David and Brian, with their heavy back-packs, loaded with equipment, padded past me and disappeared down the shadowy pathway.

'There you are,' said Jonathan. 'Nothing to it. Just take it easy. I'll be waiting at the bottom.' He swaggered nonchalantly down the almost-sheer cliff-face. Lee and I looked at each other. I knew she suffered from vertigo as well, but not in such an acute form as I did.

'Did it say anything in our contract about going down cliffs?' I enquired.

'Probably in the small print,' she said dolefully.

Offering up a small prayer, we started downwards.

There have been many times, in different parts of the world, when I have been scared, but the descent of that cliff was the most terrifying thing that I have ever undertaken. The others had strolled along the barely discernible path as if it had been a broad, flat highway and here was I, crawling on my stomach, clutching desperately at bits of grass and small plants that would, I knew, part company with the cliff-face if any pressure was brought to bear on them, inching my feet along the six-inch-wide path, trying desperately not to look down the almost sheer drop, my arms and legs trembling violently, my body bathed in sweat. It was a thoroughly despicable performance, and I was ashamed of myself, but I could do nothing about it. The fear of height is impossible to cure. When I reached the bottom, my leg muscles were trembling so violently that I had to sit down for ten minutes before I could walk. I said some harsh things about Jonathan's ancestry and suggested several – unfortunately impracticable – things that he could do to himself.

'Well, you got down here all right,' he said. 'All you've got to worry about now is getting up.'

'I shan't bother,' I said austerely. 'You may send us down a tent and arrange a supply of food parcels and we'll take up residence – the hermits of Unst.'

And in all truth it would have been a very wonderful spot to do just that. Where the so-called path ended, there was a flat area of turf, and from it one could look along the cliff-face in two directions. The shoreline was made up of a jumble of huge boulders, some the size of an average room, and among them the deep-blue ruffled sea surged and frothed and roared. As far as the eye could see along the cliff-face the rocky shoreline was alive with birds, and the air was full of them whirling like giant snowflakes above us. The cacophony of cries was tremendous. Everywhere, there were groups of guillemots sitting shoulder to shoulder on their ledges. Many of them had their single, beautifully coloured, speckled egg between their feet. Eggs green, brown, yellow, buff, spotted and blotched – like fingerprints, no two alike. Their strange, growling cries echoed among the cliffs as they jostled each other and their young. We were too late to see their courtship, but I had watched these strange rituals being enacted by guillemots elsewhere. Probably the most curious part of the courtship is when groups of birds conduct a sort of dance on the surface of the sea. They would weave and wheel, dancing over the waves, and then, suddenly, at some mysterious signal, they would all dive beneath the surface simultaneously and the dance would continue underwater. Groups of them would also indulge in extraordinary flights when a flock of perhaps a hundred birds would wheel, twist, soar and dive as if they were a single entity. What minute signals they give each other in order to achieve this extraordinary co-ordinated flight is impossible to see, but signals there must be to accomplish it in such perfect unison.

On other ledges along the cliff-face were the mud-and-root nests of the kittiwakes – neat, demure-looking gulls. While other species of gull have deserted the sea and now forage inland, following the plough and scavenging on city dumps, the conservative kittiwake has remained staunchly a seabird. It is such a delicate, self-effacing little thing that it comes as something of a shock when it opens its beak and utters the harsh cry from which it gets its pretty name. The kittiwakes of Hermaness, I noticed, were great gardeners; and hundreds of them, as they sat on their nests, were forever busy rearranging the roots and pebbles and mud that made up the cradles for their eggs.

Below the kittwakes among the rocks were the handsome razorbills with their neat black and white plumage and their beaks shaped like cut-throat razors neatly marked with white. They looked like a gathering of impeccable merchant bankers. However, occasionally one would assume what is called the ecstatic position, bill pointed skywards and then clattered like a pair of castanets, while its mate nibbled and preened its throat. This is a procedure which I personally have never seen any merchant banker (even the more convivial ones) indulge in.

On separate sections of the cliff-face the fulmars were nesting, with their dark-grey backs and tails and their white heads and breasts. There is something curiously pigeon-like about them, a look that is enhanced by their tubular nostrils. Although they are placid, even shy-looking birds, fulmars know how to protect their young. Should you venture too close to the nest, the parent birds open their beaks wide and eject a disgustingly smelly, sticky fluid from their mouths, and their accuracy is extraordinary. When I mentioned this habit to Jonathan, he was all for having me climb up to a nest so that he could film me being drenched by the parent. I said that this was definitely not in my contract and I refused to spend the rest of the day smelling like a whaling vessel. I said having my trousers decorated with skua excreta was as far as I was prepared to go in that direction.

The process of zoning on these cliffs was very apparent. At first sight, it looked like a gigantic mad concourse of birds all jumbled together, but on closer inspection you saw how neatly it was divided up. Shags had all the ground-floor apartments, then came the razorbills and the guillemots and the auks. On the higher ledges were the rows of kittiwakes and fulmars, and then the clown-like puffins on the very top of the cliffs. Among the tumbled, spray-drenched rocks, in the crevices and caves formed by the huge boulders, the shags were nesting, their greeny-black plumage shining as if polished, their green eyes as vivid as jewels in their heads. The dumpy, dark, chocolate-coloured young crouched in fear as we scrambled over the nesting-sites, but the parents abused us with harsh croaks, open beaks, glaring eyes and erect, tattered crests. It would take a brave man to put his hand into a shag's cave, I decided, for their beaks looked as sharp as knives.

On the great rocks, or 'stacks' as they are called, lying out at sea, there were cormorants nesting. Very like the shag to look at, they differed in having shiny bronze plumage and white on the chin and the cheeks. They sat upright on the rock ledges, with their wings spread out in heraldic positions. They looked exactly like the sort of thing you find on giant gateposts that guard the driveways to ancient châteaux in France. I think the cormorant, hanging itself out to dry, as it were, looks strangely prehistoric. Perhaps pterodactyls sat in that strange position.

From our vantage-point on the shoreline, we could see just opposite us an enormous stack, shaped like a giant slice of Cheddar cheese, sitting on its broad base a few hundred feet from shore. At first glance, from a distance, it did look rather like a piece of cheese covered with snow; but, when you got closer to it, it resembled much more a many-tiered and extremely untidy mantelpiece, cluttered with dozens of those horrid white pottery ornaments that you used to be able to buy with 'A Present from Bournemouth' written on them. This was the gannets' city, the white rock, on which some ten thousand gannets nested. The screeching conversation from it hit you like an almost tangible wave of sound. To say that Gannet City was busy would be an understatement. New York in the rush-hour would appear immobile in comparison. There were gannets incubating, feeding chicks, flirting, mating, preening, and launching themselves into the air in effortless flight on their six-foot wings. With their creamy-white bodies, wing-tips black as jet and their orange-coloured nape and head they were impressive and immensely handsome. Slightly waddling, slightly awkward on land, as soon as they launched themselves from the rock and slid into the air they became the most elegant and graceful of flying machines. With their long, pointed, black-tipped wings and pointed tails and dagger-shaped blue beaks, they were of a sleek and deadly design. We watched them gliding down through the blue sky with scarcely a wing-beat, using the different air currents, moving smoothly as stones on ice. They would slide up to the rock, wing-tips almost touching it, and then, turning, fold their wings and land with such a quick movement your eye was deceived. One minute they were a great white and black cross in the air, the next minute one of the multitude of noisy, restless inhabitants of the colony.

Further out to sea, we watched their incredible fishing technique in operation. They would glide along, a hundred feet or so above the waves, their pale eyes keenly watching. Suddenly, they would twist in mid-air and plummet down, their huge wings stretched out behind them so that they became a living arrowhead. They would hit the water at tremendous speed, ploughing up a bouquet of spray as they disappeared beneath the surface, to reappear a moment or so later with a fish in their beaks. When you got a shoal of fish and you had thirty or forty gannets all diving almost simultaneously, it was a breathtakingly spectacular sight.

We worked steadily all day, filming this gigantic concourse of birds, pausing only to have a much-needed picnic lunch. The weather remained brilliant, and the hot sun and the reflection on the water gave us all sunburn. In fact Lee got so red that I told her she looked like a puffin with a wig – a description which, for some reason, did not amuse her. By the time evening came, we had filmed seabirds indulging in every kind of activity and it had been an enormous privilege to spend the day within touching distance of so many species who, when they became used to your presence, completely ignored you and went about the all-important business of rearing their young, loving their mates, bickering with their neighbours in a thoroughly human fashion.

As the light started to fade and the sky turned from blue to lavender, we packed up and reluctantly left the seabird metropolis. I will draw a veil over my ascent of the cliff; suffice it to say that it was even more gruelling than the descent had been and on reaching the top I crawled across the turf as far away from the cliff's edge as possible and lay on my back, staring up into the pale evening sky while Jonathan, showing a rare Christian instinct, unearthed a bottle of Glenmorangie from his bag and plied me with it. Then we walked back over the velvety turf through the heather now purple-brown in the twilight, the cotton-grass glimmering all around us and the steady *whoosh, whoosh* of the huge dark skuas dive-bombing us in the gloaming.

It seemed incredible that in one day we had managed to obtain all the seabird footage we needed for the programme. There were only a few landscape shots we needed, which we got on the following day. Our filming on the magnificent cliffs of Hermaness was over, and so we returned to Jersey.

Here we planned to film the life on the rocky coasts. Although only nine miles by five in size, the island has such an indented coastline that you have, in effect, an enormous stretch of rocky shore for such a small area. Coupled with this is the fact that the seas around the island are comparatively unpolluted and it has a huge 34-foot tide which, when it is out, exposes acres and acres of magnificent rock-pools teeming with marine life of every conceivable sort.

The sea is a wonderful world. It is as though we had another planet joined on to this, so diverse and bizarre are its life forms, so vastly rich and colourful. From a naturalist's point of view the lip of the sea is a fascinating ecosystem where many creatures live under the most topsy-turvy conditions, several feet deep in pounding waves for some periods, dry as a bone for others. The adaptations to this strenuous sort of life are, of course, many and various. Take humble limpets, for example, so common that they are generally ignored. They have adapted perfectly to their environment. Their shells, shaped like a tent, are admirably designed to cope with the fierce pounding of the waves. The animal itself has developed a circular muscular foot with which it clings to the rock fiercely. How fiercely, you can find out by trying to dislodge a limpet with your fingers. This muscular foot forms a sort of suction cup which will enable the creature to cling so tenaciously. The limpet has evolved special gills which are like a curtain round its body. If these delicate structures were to dry out at low tide the animal would be unable to breathe and so would die. But the shell fits so beautifully to the rock that it can retain a reservoir of water to keep the gills moist until the tide's return. But it only fits so well because the limpet grinds the rock with its foot and its shell. This has two effects: a circular depression appears in the rock which fits the shell, and the shell itself is ground down to fit more closely into the rock.

When limpets graze, they move slowly over the weed-covered rocks, their small heads with a pair of tentacles protruding, and they swing their bodies from side to side. This enables the radula, the creature's tongue, to come into action. This is a strap-like organ, covered with microscopic horny teeth that rasp away the algae and the weeds. Limpets graze in a wide circle round their home depressions. It is of course vital that the animal should be able to get back to its home as the tide goes out, so that it does not become desiccated; so they have developed definite homing instincts, and how these work is still a mystery, for it does not appear to have anything to do with the creature's rather limited powers of sight, smell and touch. It's nice to think that even with a creature as common as a limpet there are still mysteries to be unravelled, that there are still enigmas for the amateur naturalist to study and perhaps solve. The limpet's sex life is confusing to all but a limpet. Like many sea-creatures, they can change their sex with comparative ease, and there is evidence that young limpets are for the most part males while the older ones are mainly females. Many limpets start life and get to be teenagers as males and then turn into females for the rest of their existence. As well as this curious state of affairs, limpets, unlike the bulk of shore snails, simply scatter their future progeny in the sea; these develop into minute, free-swimming plankton before taking life seriously and settling down on the rocks.

Limpets share this half-and-half wet-and-dry world with a host of other creatures: topshells, the woodlouse-like Triton, sand-fleas or sand-hoppers, various seaweeds and some sponges, and many of the rock- and wood-borers such as the toredo worms. But it is really in the limpid rock-pools left by the retreating tide that you find the most colourful and extraordinary creatures. Here, as well as strange methods of reproduction, you will find ingenious methods of defence and startling methods of obtaining food. Take the common starfish as an example. This beast not only, by sheer strength, pulls apart the two halves of a mussel (no mean feat, as you will realize if you have ever tried to open an oyster), but also, when the two halves of the shell are far enough apart, proceeds to extrude its stomach, push it into the shell and start the process of digestion. Then there is one of the tunicates with the lovely, slightly oriental-sounding name of Oikopleura, a tiny tadpole-like creature which has a remarkable way of obtaining nourishment. It builds out of mucus a strange plankton-trap, shaped like a minute, fat, transparent airship, in which the Oikopleura sits, wiggling its tail. This creates a current of water through the airship, which has two inhalant orifices, each one of which has a protective screen which only allows the very smallest of particles to enter. Within the airship are further mucous filters which entrap the minute organisms that make up the plankton. Also built into this mucous trap is an emergency exit through which the Oikopleura can escape when threatened by an enemy.

If methods of obtaining food are legion, so are the methods of defence and life-saving, from the sea-anemone which spits water into your eye if you prod it, to the shore crab who, if trapped by its leg, can actually amputate the leg by a muscular contraction and then grow another. The octopus, the squid and the delicate sea-hare can all produce clouds of ink to confuse and blind their enemy while they escape. The starfish can afford to lose several of its limbs in battle, as it can, in the most offhand manner, grow new ones. The scallop uses jet-propulsion to escape from a foe, shooting out a stream of water strong enough to shoot its shell several inches along the sea-bed.

Undersea sex is bewildering in its many forms. The oyster, for example, usually starts as a male, changes to a female and then, just to add to the confusion, produces both sperm and eggs alternately thereafter. The tunicate with the quaint name of Doliolum has a very complex life-history. The egg first becomes a tadpole larva and this turns into a barrel-shaped creature. Then in a certain area of the body the Doliolum starts to grow buds. These make their way down to the tail-like projection and sit there being pulled along by the parent. During this time, they gradually change into barrel-shaped beasts like the parent and finally break free.

With this rich panoply of things going on along the shore and in rock-pools, it is no wonder that Jonathan became somewhat confused. The weather, which had been so kind to us in Unst, now turned nasty. There was a cold wind and the sea, never very warm around Jersey, became icy, and there was no glimmer of sun. Every morning, we would drive down to the coast and stand shivering among the rock-pools waiting for the sun. At the slightest shift in the clouds Lee and I had to whip off our shoes and socks, roll up our trousers, seize our nets and buckets, and plunge into the icy sea.

'Try to look as if you are enjoying it,' Jonathan would bellow from the safety of the shore. 'Smile, smile.'

'I can't smile with my teeth chattering,' I would shout back. 'If you want a smile, fetch me a hot-water bottle.'

Our noses ran, our eyes watered, and from the knees down we lost all feeling in our legs and feet in the icy sea.

'Fine, fine,' Jonathan would shout. 'Now, just do that again. Go out a little bit deeper. Smile. You're enjoying it, remember.'

'I'm not enjoying it. What d'you think I am, a bloody Polar bear?'

'Never mind, the audience thinks you're enjoying it.'

'To hell with the audience.'

'You can't say that,' said Jonathan, shocked.

'I can say worse than that if you don't finish this damn sequence soon. I already feel as though I've got double pneumonia, and my wife's nose is blue and lavender like a mandrill's behind.'

'Just once more, then, and you can come out,' Jonathan would coax. The result was that Lee and I both got streaming colds, and the only compensation was that Paula who had now joined us produced several bottles of Glenmorangie and dosed us repeatedly with delicious gulps until we felt human again.

SHOOT TWO

THERE ARE few parts of the world more exposed and windswept than the lovely Shetland Islands, particularly the northernmost tip where we were operating, so it was a pleasant change on our next shoot to make our way down to the lush warm Camargue in southern France.

This is a strange area unlike anything in the rest of France – or indeed Europe. It is an area teeming with wildlife of every sort and is a breeding-ground for the white Camargue horses and the fierce little black fighting bulls. Here in the vast swamps and reed-beds nurtured by the waters from the great river there is wild boar in abundance, beaver, South American coypu, water rats and deer. Every year, tens of thousands of birds pass through on their migration from Africa to their breeding-grounds in Europe. Hundreds of species stay to breed in the Camargue itself, and so it forms a unique and important sanctuary.

In this programme, we wanted to show how important

this sort of wetland is, not just to creatures such as the wild boar that live there all the year round, but also to the creatures that pass through or that visit in order to breed. For some strange reason, all over the world man seems to think that wetlands are inimical to him. As soon as he comes across a wonderful swamp or marsh teeming with wildlife he becomes unhappy until he has covered it with pesticides, shot out all the edible animals, drained it, ploughed it, planted a series of useless crops on it and, finally, through his unbiological activities, created a sterile piece of eroded earth which was once a rich, balanced tapestry of life. This ridiculous and dangerous policy has been adopted all over the world to man's own detriment. Places like the Camargue have for millennia provided man with food in the shape of mammals, birds and fish and other fresh- and salt-water creatures, and they have provided reeds for thatching, for fencing or for firewood and a host of wild plants for flavouring and medicines. In addition, the Camargue was a sanctuary where wildlife could live and breed, and thus, if you want to look at it from this point of view, it was constantly restocking man's larder without labour and at no cost. All it took was a lack of interference. So, before the Camargue (although designated as a national park) disappears, as it undoubtedly will before the relentless pressure of what is euphemistically termed 'progress', we wanted to try to show its importance as one of the most wonderful wild areas in Europe.

Lee and I have a particularly soft spot for the Camargue, for our small house, Mas Michel, is situated in twenty-five acres of wild Garrigue, which is outside the city of Nîmes, which itself is only a twenty-five minute drive from the heart of the great wetland. Here we have eaten many memorable meals and drunk vast quantities of good wine, bathed and sunbathed, watched the crimson pink of the flamingoes like a sunset cloud, bee-eaters glistening like opals and hoopoes as pink as any salmon. Through the streets of Arles, we have watched the *gardiens* of the Camargue (the 'cowboys' of the region) ride on their white steeds with their beautiful ladies, each decked out in ancient costume. Later, they form an arrowhead of horses, surrounding a flock of the little devil-black bulls, which they drive at a gallop through the streets to the arena while the crowd does its best to break the arrowhead to make the bulls scatter, which would be a black mark against the abilities of the *gardiens* if it should be done. Then we have jostled our way into the Roman arena, tiny and beautiful, in honey-coloured stone and, when full of people, looking like a strange bowl full of moving flowers. Then the sound of *Carmen* fills the air, and the doors fling wide. From where perhaps lions or elephants and their Christian victims had at one time emerged, would appear a solitary bull, black and glossy as jet, tiny, muscular, horns showing white as ivory, strong, short legs as nimble as a ballerina's. The doors would shut behind it and he would stand there, a tiny black mark like an inkstain on the pale sand floor. He would look round, snort, trot forward a few paces, then lower his sharp horns, paw the sand, defying all comers. The fight had begun.

Before my indignant reader throws down this book and takes up a vitriol-dipped pen to write to me about the cruelty of bullfighting, may I hastily point out that there are two forms of bullfighting, and in this one the bull is never killed. Indeed, he stands an excellent chance of inflicting crippling, sometimes mortal, damage on his foes, the *razateurs*, and, as I witnessed myself, actually enjoys the fight, once he is habituated to it. The black bulls of the Camargue realise that this is a curious sort of game, yet their ferocity is such that they can kill without necessarily meaning to.

The fight is organised like this. The bull, before entering the arena, has certain little coloured tassels attached to his horns with rubber bands. These are called cockades. The object of the fight (not so much a fight as a contest of speed and skill) is for the *razateurs*, or bullfighters if you will, to remove these cockades from the bull's horns within a given length of time. No bull is in the arena for more than twenty minutes. During that time the *razateurs*, a group of men incongruously clad in white shirts and flannels as if for tennis, enter the ring. They are armed only with fleetness of foot and a curious thing that looks like a curry-comb fixed by a band round their right hand, snuggling into the palm. With this instrument they try to cut loose the cockades from the bull's horns. Each cockade is worth a certain sum of money. The longer it remains, as it were, the bull's property, the higher the price rises. Some *razateurs*, with new and idiotic bulls, leave the moment of removing the cockade until the very end of the twenty minutes so that the price is high. However, if they are fighting an experienced bull, they might find that at the end of the twenty minutes the animal still retains its cockades. The fight then ends and the Grand March from *Carmen* is played in the bull's honour as he is ushered out of the ring and back to his lorry to be taken back and released once more in the green, glinting pastures and reed-beds of the Camargue. That experienced bulls actually enjoy this I have witnessed with my own eyes, as some years ago we made a film of the Course Libre, as it is called, and so had to attend many bullfights to get the necessary shots.

An experienced bull enters the ring and gazes around at the crowd like an actor summing up the audience at a matinée. Then he goes through the 'Look how fierce I am' routine – the snorting, the tossing of the head, the pawing of the sand. He is apparently unaware that the white-flannelled *razateurs* have entered the ring and are now approaching him. Then suddenly, with astonishing speed and agility, he whirls round and is among them, galloping head down, and the *razateurs* are running before him like snowflakes driven before the wind. He chases them to the six-foot-high wooden barrier, over which they leap with alacrity – great leaps that would be envied by Nureyev himself. Sometimes the bull, to prove his fierceness, sticks his horns into the planks and sends them flying like matchwood. At other times an overenthusiastic bull might jump the barrier with the *razateurs*, and then you will see the audience in the front three rows of the stalls hastily vacate their seats until the bull can be inveigled back into the ring. On many occasions, I have seen a bull so enjoying himself that when the twenty-minute bell went as the signal for the end of the contest he would refuse to leave the ring as he wanted to continue the fight. In these cases a lead bull, with a bell on it, would be sent in to entice the recalcitrant bull out. On one never-to-be-forgotten afternoon, a lead bull was sent in and got so carried away that it started chasing the *razateurs* with the bull it was supposed to be luring out of the ring, and a third bull had to be sent in to lead the two of them out. No visit to the Camargue is complete without witnessing one of these amusing fights. Many bulls make great names for themselves, and their careers are followed by the Provençals as eagerly as if they were boxers or wrestlers or footballers and people will travel many miles to see a specially famous bull appear in the ring.

As our little house was still in the hands of plumbers, masons and carpenters, we stayed in a charming hotel in the back streets of Arles, a hotel with a beautiful tree-shaded garden in which we could sit and drink and have script-conferences. A few minutes' drive and we were out in the Camargue itself. The weather was kind to us, as it usually is in the south of France, and to get up early in the morning and know you were going to have endless sunshine all day long was very calming to our director's nerves. Our first task was to visit the various hides which had been constructed in the reserve and try to film the massive waterbird concentrations which had assembled in the marshes, some nesting, some *en route* to other breeding-grounds.

Our guide to this area was one Bob Brittan, a short, slender man, who had a mischievous urchin face and a great fund of knowledge of the Camargue, where he had lived and studied for some years. He was immediately rechristened Britannicus – a sobriquet that in a strange way suited him.

In its own way, this huge quantity of water and land birds was as impressive as anything we had seen in the Shetland Isles. Sitting in the hot hide, looking out on the acres and acres of glittering water and vivid green marshland covered with an ever-moving multitude of birds, was a great experience. Huge rafts of green-headed mallard, the rusty-headed widgeon or neat, green-eyed teal, shelducks in their carnival colours of greenish-black and rich chestnut, rusty-headed pochard, all these speckling the water or purring through the air as they wheeled in flight from one area of marsh to another. In the shallows, storks fished. Occasionally a pair would face each other, throw back their heads and rattle their red beaks like the crackle of Lilliputian musketry. Snow-white spoonbills with their strange spatula beaks, like deformed ping-pong bats, moved solemnly along, sifting the plankton-rich mud through their beaks. Flamingoes like huge pink rose petals moved along the shallows keeping up a constant garrulous ugly honking out of keeping with their elegance and beauty. Then there were the squacco herons, pale as caramel, blue and black beaks and their legs pink with the excitement of the breeding season. Soberly dressed bitterns, standing in reed-beds, doleful as bank managers contemplating their overdrafts, and rather piratical-looking night herons, with black backs and black caps and debonair, drooping white crests and red all-seeing eyes. Next to them the purple herons seemed sinuous and snake-like, with their long chestnut necks and their harsh cries, a sort of feathered Uriah Heep. Then in complete contrast were the other waders: sandpipers, pattering along the mud like schoolgirls in their first high heels; redshanks and greenshanks; the black-winged stilts with legs like all those lovely girls you see in America, whose shapely legs seem to start immediately under the chin. Then that paragon of all wading birds, the avocet, moving elegantly on stormcloud-blue legs in a black and white suit, obviously designed by the most expensive Paris fashion-house, aristocratic tip-tilted noses being occasionally dipped into the water and moved from side to side like delicate, beautiful metronomes. In the banks along the edges of the marshes, the bee-eaters skimmed in and out of their nestholes, gleaming sea-green and blue, and in the groups of pines that huddled together like crowds of furry green umbrellas the cattle egrets nested, looking like white stars in a green sky.

There was so much going on that it was difficult to know what to film. So much courtship and flirting, so much foraging for food, so much bickering and quarrelling, so many whirling flights of birds freckling the sky and splashing back on to the still waters in rose-beds of foam. Even in the hide itself your attention was constantly distracted by now a jewel-eyed wolf spider stalking a mayfly, now a butterfly hatching. In the reed-beds that flanked the hide, choking the ditches with their green stems, there were fat, ornate frogs, shiny as if enamelled, and snakes wriggling in pursuit of them. On each spear-shaped green leaf of the reeds you could see imprinted by the sun, like little black seals, the shadows of the tree-frogs. Turn the leaf over carefully, and there would be the thumbnail-size emerald-green amphibian, moist and sticky as sweets, with huge dark eyes.

The great difficulty of filming a programme like this is that your script has to be as flexible as possible. For example, I had suggested that we try to film beavers, since most people think of them as purely Canadian creatures, and do not realise that they exist in Europe. However, the difficulties of doing this were so great that our time-limits would not allow it. And so we had to substitute the more readily accessible coypu instead.

Coypu is not of course a native European species. Like so many other creatures that have become pests, like the mink, for example, the coypu was imported from the great river systems of South America to be bred in captivity for its handsome fur, sold under the name of nutria. As always seems to happen, some escaped and, finding the rivers of England and the Continent to their liking, established themselves and flourished in great numbers. They had indeed found a paradise, for in these places there were no natural enemies to keep their numbers down and so they multiplied unchecked. As coypus are large animals (a fully grown male can weigh up to twenty-seven pounds) and make extensive burrows in river and canal banks, they have become a major pest, causing floods and erosion.

Coypus are curious and rather endearing creatures, as we found out when we went down to a series of small canals in the Camargue to film them. These canals cut through the flats and are not more than thirty feet wide in places and two or three feet deep. The water is slow-moving, warmed by the sun, and the banks are lined with succulent green vegetation, so they form an ideal habitat for these giant rodents. The tamarisk trees that grew along the banks were all wearing their untidy wigs of pale dusty-pink flowers and in places groups of yellow flags made blazes of colour, looking like huge pats of butter from a distance. Around us the swallows flew *ventre-à-terre*, gleaning the mass of insect life in the clover meadows, starred with daisies, bright with pimpernels. Huge swallowtails, yellow and black tiger-striped, flew like aerial blossoms over the reed-beds and banks of flowers in the hot sun. We knew we had reached coypu country by their droppings, which were in great profusion floating in the slowly moving canal water and littering the bank, and very distinctive droppings they are, too, each some two and a half inches long, resembling short, blunt cigars in shape and colour. They are finely ribbed like some beetles' wing-cases from stem to stern.

We frequently had to cross and recross the canals using primitive bridges which, for the most part, were rotten logs or planks jumbled together and flung haphazardly across the water, and all highly unsafe. So we had to form human chains to make sure the precious and extremely expensive cameras and recording equipment got over safely. This was accomplished with no little noise, although we tried to be as quiet as we could. However, when we arrived at our destination, the coypus, as was inevitable, had heard us and disappeared.

'Damn them,' said Jonathan. 'What do we do now?'

'Wait,' I said succinctly.

'We're losing good filming-time,' Jonathan complained.

'You're filming wild animals, dear boy,' I explained, not for the first time, 'not film stars. Animals don't take direction.'

'What about Lassie and Rin Tin Tin?' he retorted.

'Hollywood products,' I said. 'You'll just have to possess your soul in patience. Look at these very fascinating droppings.'

'I can't make a half-hour programme about a coypu's droppings,' said Jonathan with, I must admit, some justification.

'Patience,' I said soothingly. 'They will return.'

But I was wrong. They did not return, and after several hours during which I tried to quieten Jonathan down with a recital of all the poetry and limericks I knew and reminiscences of similar failures in animal photography I had known in my life, none of which appeared to have a calming effect, we decided to take Britannicus's advice and come back in the evening, when the coypus would emerge to feed – or so he assured us.

So we went away and filmed birds and returned in the late afternoon. Knowing the paths and the bridges, we made better progress and less noise, and were soon ensconced in the tamarisk trees, which grew so thickly that they made an admirable hide. If the coypus appeared, we decided, we would get all the shots of them we needed first and then Lee and I would see how closely we could approach them, for Jonathan was anxious to get a scene with both the talent and the animals in the same shot.

'I am sick to death of those animal series that show the talent peering through binoculars, creeping through the undergrowth, and then you cut to a shot of a penguin doing a Highland fling,' he confided in a hoarse whisper, 'and you know perfectly well that the talent probably never even saw the penguin.'

'They'd be jolly lucky talent if they saw a penguin doing that,' I said judiciously, 'but I do know what you mean.'

'What I mean . . .' Jonathan started, and Lee shushed him.

'I think I can see something black in the water over there,' she said, pointing.

'Probably another dropping,' said Jonathan mournfully. We all peered hopefully at the canal and then saw a blunt, bewhiskered head with ridiculously huge yellow teeth break the water and move slowly along the surface, leaving a V-shaped trail of ripples behind it. Frantically, we tried to attract Chris's attention, for he was some distance away, but he had already seen it and was busy filming.

The coypu's head reached the bank and the portly animal hauled itself ponderously ashore, displaying a behind of gigantic proportions, like a fat, fur-covered balloon. It had great naked flat feet and a long, thick scaly tail like a rat's. It sat up on its ample backside and sniffed the air suspiciously, its front feet bunched into absurd fists, its enormous protuberant yellow teeth making it look as though it was grinning. All it needed, you felt, was a monocle and an old school tie, and it would be what the average American thinks the average Englishman looks like. Satisfied that there was no danger, it proceeded to groom itself carefully, using its front feet. The coypu has two sebaceous oil-glands situated at the corner of its mouth and near its anus. The fur consists of a thick, rather harsh outer coat and a thin, fine undercoat. When the nutria is used commercially, the harsh outer coat is removed, leaving only the soft undercoat. We were amused to see how assiduously the animal groomed and arranged and oiled its fur, taking immense care and concentrating intently. While it was doing this, several other coypus' heads broke surface, and these other animals of varying sizes from quite young ones to big fat matrons were hauling themselves out on to the bank. Soon there were half a dozen of them mostly sitting on the bank and grooming, while others swam and dived in the canal. As each one finished grooming, it would meander along the bank browsing on the succulent vegetation. They seemed charming, placid and nice-natured beasts, an ornament to any scenery, if only they would stop carrying on like a corps of engineers, undermining every bank they came to.

Presently, Chris signalled us that he had obtained all the necessary shots, and suggested in dumb show that Lee and I should approach the disporting coypus to get the shots Jonathan wanted. The evening was so still, we did not have to worry about wind direction. All we had to worry about was not letting our heads show against the skyline. Bent double, like a couple of Red Indian trackers, Lee and I made our way along the canal. We arrived at the tamarisk tree with one broken branch dangling from it that we were using as a marker. We were now opposite the coypu colony. Very slowly we rose to our feet, inch by inch, with frequent pauses. Then we stood upright, and twenty-five feet or so away were the coypus.

We stood very still and they seemed unaware of our presence, and went on with the grooming and bathing. Slowly we inched forward. It was rather like playing that ridiculous childhood game of Statues or Grandmother's Footsteps, where a group of you approach a person with his back to you and when he suddenly turns round, you are all supposed to freeze. At the slightest movement, you become 'it'. So Lee and I played Grandmother's Footsteps with the coypus, and by this means succeeded in getting quite close to them, and Chris managed to get both them and us in the same shot. We were standing frozen when the first male who had appeared suddenly turned round and stared across the canal at us, nose whiffling, whiskers bristling, orange teeth like scimitars pointing at us. As we were immobile, we could not see what had alarmed him, but it may have been a tiny breeze from the wrong direction wafting our scent to him. At any rate, he suddenly squatted down on all fours and then ran purposefully down the bank and entered the water, making hardly a ripple for such a large animal. The next minute, panic ensued among all the others and they galloped down and leapt into the canal, churning up the waters as they dived beneath the surface for safety.

Later that evening as we sat over drinks in the green twilight under the plane trees in the hotel garden, Jonathan was smug and satisfied after the successful filming.

'That was a good day's work,' he said. 'Now we've only got the pigs and the bulls to do. How's Pig Woman?'

'She'll be there at sunset tomorrow,' said Britannicus. 'I should bring lots of mosquito repellent if I were you.'

'Oh God, not mosquitoes,' said Brian, rolling his eyes. 'You know how they love me.'

'Highest concentration in Europe, I should think,' said Britannicus mischievously. Brian groaned.

'I don't know why everybody makes such a fuss about a few

mosquitoes,' said Jonathan airily, 'they never worry me.'

'I shouldn't think any self-respecting mosquito would want to bite you,' said Chris judiciously, showing the usual cameraman's love and respect for the director.

'Who is Pig Woman?' enquired Paula, who, having succumbed to the rich food and wines of La Belle France and not having a cast-iron stomach like the rest of us, had wisely spent the day in bed, and so had not caught up with events.

'Pig Woman,' explained Jonathan with relish, 'is a young zoology student who is studying the wild boars of the Camargue. She traps them, puts radio collars on them and then rushes about in a van plotting their movements. This we will film.'

'Couldn't she do it in the day when there are no mosquitoes about?' asked Brian hopefully.

'Pigs don't move around in the day much,' said Jonathan, 'so I understand. True, Britannicus?'

'Yes,' said Britannicus, 'to a large extent *Sus scrofa* feed at night, particularly in the areas where they are liable to be hunted during the day. They are actually being hunted now.'

'Poor things,' said Lee indignantly. 'Why do they have to hunt them?'

'Well, when you think of the damage they can do to crops by their rooting, plus the fact that if the food has been abundant in one year they can have two litters with up to six young, the farmers feel that they must keep the animals under some sort of control,' explained Britannicus.

'Not to overlook the fact that wild boar meat is considered a delicacy,' I said.

'True,' said Britannicus, grinning. 'I am sure there are parts of the Camargue where the damage they do is exaggerated as an excuse to hunt them.'

In the late afternoon, we drove out to meet Pig Woman. The dirt roads ran straight as rulers, white with the salt crust on them, between fields of mauve sea-lavender which, from a distance, looked like pale smoke drifting in a carpet two feet high over the ground. Here and there were thick patches of false olive, a silvery-green tree growing some six feet in height and very bushy. The young trees looked from a distance tight and curly as though someone had practised a curious form of topiary on them. After several miles, the false olives grew thicker and we eventually arrived at a glittering white crossroads among huge thickets. The blue sky had a faintly gold wash to it and a few pale clouds like tiny feathers hung immobile in the western horizon, turning from white to gold and then to pink. We parked the cars and waited for Pig Woman. Presently, she arrived, bumping over the dirt roads in a tiny, battered Deux Chevaux van with a long antenna sticking out of the roof, which wriggled and whipped like a fishing-rod with an infuriated marlin on the end. She chugged to a halt, and stopped the van and got out and walked towards us. I don't know why, but the sobriquet 'Pig Woman' had conjured up in my mind something out of a horror story, some snouted, grunting half-pig, half-woman, with huge tusks and slavering jaws and doubtless unpleasant habits like eating its young. So in consequence I was somewhat relieved to find myself being introduced to a slim and handsome young woman who had none of the less attractive attributes of a member of the Suidae. Marise was her name, and she viewed us with a bright humorous eye while Jonathan explained what he wanted. She obviously thought us quite mad but was happy to oblige the eccentric *anglais*. First, as the sun was setting, Jonathan wanted her to drive her truck with its whipping antenna to and fro through the false olive groves, just as she would do if tracking the pigs after dark. These shots had to be done before the sun set, so that when printed in the laboratory they would look like night. This she dutifully did, and by the time we had finished it was almost dark.

The mosquitoes, as though at a signal, rose from the surrounding countryside and converged on us like a solid wall. I have always contended that no place on earth could compete with the Paraguayan chaco near the Matto Grosso in quantities of mosquitoes. After our experience in the Camargue, I am hesitant to give Paraguay first prize. Wherever you shone your torch all you saw was a thick, dancing, almost opaque veil of mosquitoes. It became advisable to breathe through the nose if you didn't want to inhale a lungful of them. Our hands and faces and necks became black with them. They bit our scalps through our hair and they bit every other portion of our anatomies through the thin summer clothes we were wearing. In seconds, Brian was whirling like a dervish, slapping and, moaning. Reeking though he was of so-called mosquito repellent, this made no difference. The mosquitoes of the Camargue apparently looked upon the foul-smelling repellent as a sort of aperitif before getting down to the main meal of blood. Lee and I knew it was more than our life was worth to point out that mosquitoes did not worry us. Of course, they were irritating when they flew in your eye or up your nose, but because of Lee's two-year research in Madagascar and my wanderings about the world we have developed hides like rhinos and few if any of the bites we suffered even itched. But you can't tell the true sufferer that without risking a lynching.

While the team, cursing and swatting, set up the lights for the next scene, Marise, Lee and I shared the back of her little van with approximately two million mosquitoes as she talked about her studies. In the old days, if you wanted to work out where animals went and what they did you had to rely on your own eyesight and tracking abilities. Now, with radio tracking, the whole thing is more efficient and accurate. The small radio collars attached to an animal send out a radio bleep. This is picked up on what, to all intents and purposes, is a little radar screen. Reading the bleeps off the screen on to a map of a given area, you can follow the movements of the animals you are studying without having to disturb them or go near them. Marise was obviously very involved with the wild boars she was studying and she talked on enthusiastically, oblivious of mosquitoes. Did we know they had a tremendously wide range of food? Even though the bulk of their food is vegetarian – acorns, beech mast, grasses, various herbs – they will eat an astonishing amount of other things from carrion, ground-nesting birds and their eggs and young, lizards and snakes, various insects, crabs, and they have even been seen to be adept at catching mice. During the rutting season, she went on, the big boars have the most terrible mating battles, slashing at each other with their razor-sharp tusks. Generally, they tend to attack each other's shoulders, and to protect himself against his sharp-toothed foes each boar develops at this season a sort of thick plate of flesh which guards his shoulders much in the way that knights of old used to wear steel breastplates for battle. When the female is ready to farrow, she leaves the others and finds a quiet place in dense cover where she builds a comfortable nest, sometimes even roofing it over, and here she gives birth to her piglets.

By now the lights had been set up, illuminating the interior of the mosquito-ridden van and Marise's equipment, and we were ready to show how it worked. She prepared her maps and switched on her little radar screen, and then she started slowly to turn the fishing-rod antenna on top of the van. Soon there was one bleep on the screen, a tiny green dot; then there was another and another and finally a little constellation of them. It was a fascinating thought that we, sitting in the van perhaps a mile away from the pigs, were aware of the movements of these elusive and wary creatures, and yet they were unaware of our surveillance. Eventually, thoroughly bitten but happy, we thanked Marise for her help and, leaving her to her work, went back to our hotel, having arranged to rendezvous early the following morning to visit her traps with her.

The eastern sky was just paling into daffodil yellow as we drove out along the straight white roads and into the dark, mysterious, more solid thickets of false olives. There was much birdsong, and skeins of duck flew black against the lightening sky, flighting into the deeper recesses of the marsh to feed. Presently we stopped the cars and got out, and walked a few hundred yards to a clearing where Marise had her trap – a huge box made out of wood and wire and steel, baited with all sorts of delicacies. It was important that the trap was looked at for results at first light, for if you left the boars in the traps when the sun had risen you risked losing them through sunstroke. It is always interesting to visit a trap, wondering whether your efforts have been successful. I have set traplines all over the world, but I have never got over the excitement of visiting traps in the early morning, wondering what you have caught, if anything. In this case, we were well rewarded, for in the trap was a whole sounder of baby boars, six of them, each about the size of a terrier, with gingery fur, and their baby stripes just fading. To Marise's delight, one of them was wearing one of her radio collars. He had apparently led all his brothers and sisters back to the same trap in which he had himself been caught.

The babies jostled together in the trap, stamping their feet, grunting and squealing, obviously a bit panicky at our approach. Marise and her team of helpers worked very swiftly. The job had to be done as quickly as possible, so as to minimise the stress on the piglets. Each baby was ushered from the main trap into a sort of funnel trap from which, with many piercing screams, he or she was gently extracted and a radio collar deftly clipped round its neck; the next minute it was released and making off through the trees at a smart trot, tail twisted up over its back in an indignant question mark, carrying with it unknowingly the instrument that would allow us access to its private life. It was, I reflected, rather unfair. It was like having the local police station bug your bedroom. However, if we are adequately to protect wildlife we must know how it functions and what its needs are, and this is one of the many ways of doing just that.

The next morning, with his mouth full of croissant, Jonathan said jubilantly and indistinctly: 'I've fixed up the bulls.'

'Good,' I said absentmindedly. 'What bulls?'

'Well, you know you said you couldn't show the Camargue without showing bulls, so, I've fixed up some bulls.'

'But they're not fighting at this time of the year,' I pointed out.

'I don't mean fighting,' said Jonathan. 'I mean we're going to round them up.'

'This use of the royal We,' I said cautiously, 'does this include Lee and me?'

'Of course you,' said Jonathan, with the air of one promising a treat to a child. 'You'll go off into the swamps, round up this great herd of bulls and drive them past the camera.'

'What do you mean, "drive them past the camera"?' I asked. 'These are bulls, not dairy cattle.'

'You'll be all right, you'll be on horseback,' said Jonathan.

'Oh, what fun,' said Lee eagerly.

'It's not going to be fun at all,' I assured her. 'Do you realise, I haven't been on a horse for nearly thirty years, and you expect me to gallop about rounding up fighting bulls.'

'You'll be all right,' said Jonathan. 'It's as easy as—'

'Don't,' I interrupted. 'You used that metaphor about the cliff in Unst, I remember, and it was anything but as easy as falling off a log.'

'They can give us some very old horses,' suggested Lee helpfully.

'As far as I'm concerned, the only horse I'm going to mount has got to be more than ready for the knacker's yard,' I said.

'It's all right,' soothed Jonathan. 'They've promised they'll choose very gentle horses.'

'You'd never do this to the talent if Paula were not confined to bed,' I said. 'She knows how to cosset the stars.'

As it turned out, the horses they had chosen were massive, benign beasts who obeyed our every instruction, and the saddles, being built on the American pattern, were as comfortable and as well padded as armchairs and about as difficult to fall out of. The cameras were taken out to the edge of the swamps and set up among some tamarisk groves; and then, accompanied by about ten or twelve gypsy-like *gardiens*, we set off into the swamps in search of the bulls.

When you are in practice, travelling on horseback is one of the finest forms of travel for a naturalist. You can go as fast or as slowly as you like. You can stop to observe without necessarily dismounting, and your steed, moreover, will allow you to visit areas no other form of transport can penetrate. In addition, you have the added bonus that wildlife on the whole treats a human on a horse as less of a menace than a man on foot.

So we set off, the sun hot on our backs, the sky pimpernel blue above us, through the pink and green tamarisk, our horses' hoofs splashing through the six inches of limpid water that covered the lush grass and reeds. Here and there were beds of flags, glinting yellow as gold in the sun. As we advanced into the swamp the water got slightly deeper and each time our horses' hoofs splashed into it they flung up a fountain of droplets and each one was taken by the sun and turned for an instant into a miniature rainbow-coloured planet, whirling through the air. Burnished frogs slid under the water away from the monstrous hoofs, and around us darted huge blue and huntsman-red dragonflies. Flocks of the more delicate damsel flies in pale powder blue and deep peacock blue rose from the clumps of flags as we splashed through. Once a huge scarlet dragonfly sped past on purring, glittering wings carrying in its jaws a bright-blue damsel fly. Around us, opalescent bee-eaters and dark swallows hawked the myriad insects, and in the distance we could see herons, bitterns, egrets and night herons pursuing frogs and tiny fish among the tamarisks.

Suddenly, ahead of us we saw the bulls. A herd of about a hundred animals, grazing beneath the trees, looking like a black and dangerous reef against the green of the swamp. The *gardiens*, telling us to go to one side for a moment, then spread out and surrounded the snorting, suspicious creatures, whistling and calling encouragement. Gradually, they moved the bulls towards and past us, and we then took up our position behind the herd, jockeying them along. At first the bulls moved slowly but then, encouraged by the *gardiens*, they started to trot and then the whole black mass, horns glinting, broke into a gallop in a great froth of water and we galloped behind them. It was really most exhilarating with the herd thundering along on a tidal wave of water and we riding behind, shouting and whistling in imitation of the *gardiens*.

Then, quite suddenly, it ceased to be exhilarating.

The bulls reached a rather solid thicket of tamarisk trees and for some reason best known to themselves they decided that danger lurked beyond the trees. The herd halted and turned and, as one animal, came thundering down on us. One minute we had been gaily pursuing the bulls, the next minute we had turned and were in full flight. The great mass of menacing black muscle surmounted by a forest of sharply curved horns thundered after us. It was a very confused and hectic five minutes before the *gardiens* managed to turn the stampede. They let the bulls graze for a few minutes so that they could regain their equilibrium (to say nothing of us regaining ours) and then slightly less vigorously we chivvied them towards the cameras.

It was at this moment that I got my own back on Jonathan. A hundred yards away, he and the camera were not very well concealed behind a group of fragile tamarisks. The bulls took fright again. This time they were convinced that the danger lay behind so they broke into a gallop and thundered down upon the cameras. The idea had been that the bulls were to be gently chivvied past as they were filmed, and now before the *gardiens* could do anything sensible a great torrent of bulls like a solid black avalanche, splintering and knocking down quite big tamarisk trees in their panic, engulfed both Jonathan and the cameras. Fortunately, the bulls seemed too panic-stricken to notice either Jonathan or Chris and passed by them on either side, taking most of the tamarisk grove with them.

I rode up to where Jonathan and Chris stood looking considerably shaken.

'What ho, Harris,' I said jovially. 'Wasn't that fun?'

'Fun?' he said hoarsely. 'I thought we were done for. I've never been so scared in all my life.'

'You do fuss,' I said airily. 'It was only a few bulls.'

'A few bulls,' said Jonathan indignantly. 'There must have been hundreds of them. We might have been killed.'

'Well, I don't know what you're worrying about,' I said. 'It turned out exactly as you said it would.'

'What do you mean?' asked Jonathan suspiciously, mopping his brow.

'Well, you said it would be as easy as falling off a log,' I said blandly, 'and it was.'

Harris gave me the sort of look that directors give to the talent when they get out of hand, the sort of look the late Boris Karloff made famous.

SHOOT THREE

EVEN THOUGH it gets hot in the Camargue and wildlife abounds in its great swamps, it is not nearly as rich as the real tropics are, both in plant and animal life, and it was this richness we wanted to show. The northern and southern areas of the American continent are nipped in the middle like an egg-timer by the isthmus of Panama, and through this narrow pigtail of land the tropics of Brazil escape from the southern half of the continent and spill out into Ecuador, Honduras and Mexico, and then gradually fade northwards into the more temperate regions of the United States.

Panama is a fantastic country, the sort of country of which a naturalist dreams, for here he can explore the indescribable richness and complexity of the rainforest in the morning and in the afternoon be swimming off an immense colourful reef, teeming with life. It was for this reason that we chose Panama, for the dreaded budget would not allow us to go traipsing all over the world, and so in a small country we had conveniently at hand both forest and sea. We wanted to try to show how alike in many ways the structure of a coral reef and that of a tropical forest are, for if you substitute coral and weed for tree, fish, crayfish and other sea-creatures for birds, mammals and reptiles of the forest you are astonished how similar the two ecosystems are.

Panama had another advantage as far as we were concerned: ever since the construction of the canal and the necessary flooding that went with it, an island was created, called Barro Colorado, and it has been used for many years by the Smithsonian Institution as a tropical research station. The Smithsonian also has the reef research station on the San Blas Islands, lying off the Caribbean coast, about an hour's flight from Panama City. Wherever groups of scientists gather together in one spot over a period of time, you may be sure that they get to know every leaf of every tree, and this sort of knowledge, when your time is limited, is of inestimable value to the film-maker.

Lee and I arrived in Panama City suffering terribly from jetlag, since we had flown the Atlantic and then down to Panama from New York. However, no exhaustion could quench our happiness at being in the tropics again, to see the boat-tailed grackles, black and solemn as undertakers, parading on the half-finished blocks of flats outside our hotel bedroom window, to see glittering humming-birds and butterflies the size of your hand in the hotel garden and, above all, to feel the moist, scented hot air, like the smell of plum cake from a newly opened oven, that told you that you were once more in that richest area of the earth's surface, the tropics.

The following day, when we had recovered, we met for a briefing with Paula and Alastair. Alastair has a very curious method of communication with members of his own species. So strange is it, indeed, that I, in spite of priding myself on being able to communicate with most people anywhere in the world, found I had to use Paula as a translator. What Alastair would do was to throw you a half-sentence or, even worse, two half-sentences which appeared to have no connection with each other, and you then had to fill in the missing words to find the sense of what he was saying. It was rather like trying to do a *Times* crossword puzzle without the clues. Now, beaming at us affectionately, he said: 'Jetlag over? Good. I thought . . . you know . . . San Blas first. Reefs like . . . or perhaps more like . . . forests, fish really, like birds only no wings. Don't you think? So islands . . . pretty . . . because you don't . . . see when we get there. Then we know for, er, Barro Colorado, don't we?'

I took a deep draught of my drink. It had been several months since I had worked with Alastair, and mercifully time had dulled some of the wounds brought about by the more horrifying attempts at communication with him in Mauritius. I threw a mute look of appeal at Paula.

'What Alastair is saying, honey, is this,' she said soothingly. 'If we are going to try to compare the forest with the reef, he thinks the reef is going to be more difficult because it is underwater filming, so he suggests we go to the San Blas Islands first. OK?'

'OK,' I said. 'I don't mind.'

'OK, so we leave tomorrow. Is that all right by you guys?'

'Sure,' said Lee, and then made the mistake of trying to extract further information from our director. 'What are the San Blas Islands like?'

'Covered in . . . you know . . . pretty things, palms, that is islands . . . um, many of them Indians, government can't control . . . women . . . gold in nose, so forth. Reef, big ones,' said Alastair, waving his arms excitedly. 'You'll like it . . . sure to . . . Conrad.'

'Haven't you got a book on them?' Lee asked Paula hopefully. As a travel guide, Alastair was obviously not going to be terribly coherent, though obviously enthusiastic.

I have often thought that if Martians ever landed it would be just their luck to run into this kindest, most liberal but most incomprehensible of men as their first example of the human race.

So the next day we assembled early in the morning at a tiny airfield on the edge of the city. Our cameraman for this shoot was Roger Moride, a tall, handsome Frenchman who looked and sounded like the late Maurice Chevalier. He had a great fund of amusing stories and an avuncular eye for the ladies.

We piled our odd assortment of gear into an aircraft designed to hold twelve people and when we had taken our seats we were joined by some fine-looking Indians, stocky, coffee-brown, with very Mongolian-looking features. The men wore shirts and trousers and floppy hats, but the women wore brightly coloured skirts, headscarves and blouses that had been made vivid in reverse appliqué, most beautifully constructed. One elderly lady had a large, flamboyant toucan on her chest, with a roguish look in its eye; another had two huge red fish beaming at each other, face to face in an ultramarine sea; and a third lady's bosom was covered in a spirited scene of several small black fishermen in a canoe trying, with a most fragile and ineffectual fishing-rod, to catch a school of fish the size of sperm whales. All the ladies, gay and gaudy as parakeets in this charming finery, had an additional ornament, gold rings, like wedding rings, implanted through the centre of their noses, and their cheeks were gorgeously made up with cyclamen-pink rouge. These were some of the San Blas islanders and they looked simply splendid.

We had a mildly bumpy flight over the centre of Panama and soon we reached the Caribbean coastline and were flying over blue, translucent seas with reefs showing like strange sea-serpents embedded in blue amber. Scattered all around were the hundreds and hundreds of San Blas Islands, each so small and perfect with its wedding-ring of reef around white beaches and shaggy wigs of palm that they looked like manufactured South Sea islands in a toyshop window. Presently, somewhat to my consternation, the pilot flew over the blue waters, dropping lower and lower, and headed for an island of such microscopic dimensions that it seemed impossible that he intended to land on it, except in the direst emergency. By now we were almost skimming the surface of the water, and poor Alastair, who did not like small planes any more than I like heights, was looking distinctly apprehensive. Just as we all thought a crash landing in the sea was unavoidable, we flew over a snow-white beach and immediately beyond it the tarmac started. We touched down in a series of juddering bounces and were then tearing along the tarmac, brakes screaming. It was obvious, when we finally drew to a halt, why this method of landing was necessary. The runway exactly fitted the island, so to speak, or the island fitted the runway, no room for error. If you didn't get it exactly right, you landed at one end of the runway and ended up in the sea at the other, and I don't think Alastair was the only one who was glad to quit the plane.

We waited some time after our plane had landed, our mountain of luggage smouldering in the sun, covered with brown and green grasshoppers, who appeared to find it irresistible. All our fellow-passengers had been met by canoes and were now dots on the sparkling sea, making towards the scattering of islands across the horizon. Presently, a large, deep-bedded canoe hove into sight and when it pulled up at the jetty out got a stocky little man with bow legs who looked so Tibetan you would have thought he had come straight from Lhasa. He was, it turned out, Israel, the owner of the hotel in which we were going to stay.

The shallow sea was blood-hot and as clear as gin, with small flocks of multicoloured fish flipping and trembling over the sandy bottom. We pushed off and presently we were paddling over the still waters towards an island that looked as though it might be about four or five acres in extent, thick with palm trees. We rounded a point and then headed towards a small cement jetty, behind which lay the hotel, an edifice which took my breath away.

'Look at it,' said Lee in delight. 'Isn't it wonderful? I've never seen anything like it.'

'The most extraordinary hotel I've ever seen anywhere in the world,' I said. 'Full marks, Alastair. We're going to enjoy this.'

'It's fun, isn't it?' said Alastair, beaming. He was pretty reliable on short sentences.

The hotel was charming. Shaped like a capital L, it was two storeys high, with a palm-thatched roof, and the entire building was made from bamboos lashed intricately together with a sort of raffia. A double veranda ran the full length of the L, and from it on the ground floor and the first floor doorways led into what we presumed were bedrooms. The whole thing was perched over a deep cement pool in which a myriad of coloured fish swam, accompanied by two portly turtles. Next to the hotel was another lopsided bamboo-and-palm-leaf structure with a battered sign saying 'Bar'. Interspersed with all this were tall palm trees curved like bows, rubbing their dark-green leaves together, whispering to the breeze. A riot of hibiscus and other tropical bushes was in full flower. The whole thing in the most brilliant sunshine had an air of unreality. It looked exactly like a Hollywood film-set for a great South Sea epic. One expected (and looked for in vain) a sour-faced Somerset Maugham in immaculate white ducks descending the rickety bamboo stairs. But the closest you got to it were the two turtles whose expressions of disdain were remarkably similar.

Our bedroom was, to say the least, a novelty. There was no need for windows since the light streamed through the bedroom walls, and some of the cracks afforded us excellent views over the sea and the islands around us. The beds were enormous and sagged in the middle, having at one time, it was clear, suffered a severe prolapse. The sand on the floor scrunched pleasantly under our feet and gave a special outdoor feeling to the apartment. A small cubicle the size of a coffin led off this honeymoon suite, constructed of beaten-out kerosene-tins covered with peeling oilcloth in a striking tartan pattern. From this Scottish ensemble protruded a small pipe from which, experiment proved, when a tap was turned, a jet of sea-water hit you straight in the eye. It was not the Ritz, but you didn't expect it in these idyllic surroundings.

We had scarcely unpacked and hung our clothes neatly on the only chair, when looking over the veranda rail we saw a canoe arriving, piloted by a splendidly golden-brown young man accompanied by a young blonde. This proved to be Mark, doing special research on fish at the Smithsonian Institution's research station, a clutter of buildings perched uncomfortably on a reef a quarter of a mile away. Mark had been seconded to us to be our guide and adviser while we were there. He was very attractive, with a slightly oriental cast to his features, and I found out later that his mother was Japanese. Extremely knowledgeable and competent, he immediately became our mentor and friend, as did the student Kathy who was working with him. That afternoon, Mark took us out to the reef a mile or so away, where he was doing his researches, and so knew practically every fish by its Christian name. We anchored the boat in six feet of water over the sandy bottom on the edge of the reef, donned our masks and dropped into the warm water.

I can never get over the wonder of that moment when you enter the water and find your face beneath the diamond-bright surface of a tropical sea. The mask is like a magic door, whose opening smooths out the ruffles and pleats of the water, and you slide effortlessly through a fairyland of unimaginable beauty. At first we drifted over the golden sand, patterned with its bright, ever-moving chain mail created by the brilliant sun, and saw the stingrays like strange mottled frying-pans glide out of our way. Here and there were small islands of coral, smouldering like great jewels, clad in multicoloured weeds, decorated with sponges and sea-squirts in vivid colours, each island with its retinue of fish – orange, scarlet, blue as a midnight summer sky, yellow as a dandelion, striped, speckled, pleated and ruffed, shapes to defy the imagination. We swam on and presently the reef loomed ahead, an extraordinary area of grottoes, channels, hidden gardens of sponges and intricate corals, great castles of coral with banners of weeds flying from their battlements. There were brain coral like enormous craniums of giants felled in battle who had fallen into the sea and whose bone structure had become part of the reef. All around you there was the click, purr, rasp or squeak of fishy conversation, abuse and feeding. Take one of the winding channels and follow its eel-like wrigglings. One minute the weed would be brushing your shoulders on each side, sea-urchins like marine horse-chestnut husks clinging to the multicoloured walls, the fish darting ahead of you as if enticing you on. The next minute, the narrow channel would suddenly open out into a small area of dazzling filigreed sand covered with the obese black sea-slugs as though some marine delicatessen's delivery-van had dropped a load of *saucisson* by mistake. Then the narrow channel would become a great fish-filled valley and you could feel the pulse of the sea, the lift and fall, as you drifted like a bird out over the edge of the reef and suddenly below you was nothing but mysterious, menacing blackness as the reef-edge slipped down into the sea's bottom and disappeared into velvety darkness.

Mark, over this and other reefs, had the intimate knowledge a man has of his back garden. He would tell you to swim down a certain channel, take the first left, second right, turn left at the big brain coral and twenty feet down the channel you would find the sponge or the coral or the fish that you were looking for. He directed you around the reef as a man would direct you round his home town, and certainly, without his guidance and expertise, there is a lot we would have missed or failed to understand unless it was explained. With birds and mammals and, to a certain extent, reptiles, their language consists to a very large extent of minute gestures and postures, and it takes time for you to adjust to these movements – to be able to interpret what a wolf's tail is saying to you, for example. Now, with life under the sea you have to learn a whole new language. You are constantly asking yourself why is that fish lying on its side? Or standing on its head? What was that one so busily defending and why was that one, like some street-walker, apparently soliciting a fish of a different species? Without Mark's help, we would have had little hope of understanding a millionth part of what we saw.

Take the damsel fish, for example. These plump, velvety-black little creatures are ardent gardeners. Each had selected a certain portion of coral on which grew a mass of weed, carefully tended by the fish, representing not only his own territory but his larder as well. This garden he would defend against all comers, and his bravery was considerable. One that we watched and eventually filmed had a green garden some six by twelve inches in size on a huge brain coral. Our attention was drawn to him because he was, unaccountably, and with the utmost vigour, attacking a sea-urchin as black and spiny as a pin-cushion perambulating innocently past. Closer inspection, however, revealed that the sea-urchin's peregrinations were going to take it, as it were, bulldozing its way straight across the damsel fish's front lawn, hence his display of pugnacity. One morning we found our damsel fish nearly frantic, for his precious garden was being visited by a group of parrot fish. These large, gaudy, green, blue and red fish with their parrot-like mouths swagger over the reef like groups of multicoloured muggers, and the sound of their sharp beaks rasping at the coral can be heard a surprising distance away. There were so many of them that our poor little damsel fish did not really know which one to attack first. They also had a strategy. One would zoom into the garden and rip up a piece of weed and the damsel fish would immediately attack and drive it off, although it was twenty times his size. But while he was busy chasing that one the rest would descend on the garden. The damsel fish would eventually return and scatter them and the whole process would then be repeated. Luckily we arrived before the parrot fish had done too much damage and we frightened them off. Nevertheless, in spite of the aid we had rendered him, our damsel fish never really trusted us. He suspected Lee of living on an exclusive diet of seaweed and felt sure she had designs on his garden, and so he would attack her vigorously if she got too close.

Among the many fascinating aspects of reef life Mark showed to us, none was more intriguing and bewildering than the sex life of the blue-headed wrasse. If Freud thought that the sex life of the average human was complex, he would have had a nervous breakdown if he had been forced to psycho-analyse the blue-headed wrasse. To begin with, he would have been in some doubt as to whether he was addressing Mr or Mrs Wrasse, and this alone may have given him pause for thought.

When the blue-headed wrasse are young, they are not blue-headed. It is no good beating about the bush, I might as well make a clean breast of it, they are yellow and don't even really look like blue-headed wrasse. However, don't despair. When they grow up, they undergo a startling colour-change and become deep, velvety blue with a light blue head. The male then stakes out a territory in the mountain ranges of coral and defends it against all comers and waits for the ladies. He is large, glamorous, and he can mate with as many as a hundred females a day – a fact that makes the prowess of all legendary human lovers pale into insignificance. The females, dazzled by his brio, find him irresistible and visit his coral apartment by the dozen. However, this is where the difficulties arise. Young males, too young to be able to obtain and defend a bachelor pad, hang around the adult fish's territory, waiting for the ladies. Groups of them then force the female to rocket skywards in the water and release her eggs, while the young males release their sperm and fertilize them. However, this is really unsatisfactory and obviously it is a fairly hit-and-miss affair. Ideally, the young males should stake out and defend territory and in this way be able to have the females to themselves, and thus fertilize even more. So his strategy is to grow big enough, change colour and get himself a penthouse.

Meanwhile, what of the female wrasse? It is obvious that the number of eggs she can lay and the number of offspring she can produce is very small compared to the legion a big male can fertilize. So what does she do? It sounds magical to us, but it is commonplace to a wrasse. She simply changes sex – from a yellow female into a large blue male, strong enough to seize and defend a territory. This she does and is soon busy mating with dozens of females a day. This is, I suppose, the ultimate, a sub-aqua piscatorial women's liberation movement. Love in the wrasse's world is enchanting but apt to be confusing to the amateur naturalist at first.

We arranged to film the damsel fish defending his garden and the extraordinary sexual activities of the blue-headed wrasse and many other things besides. Once Alastair got so carried away that he attempted to give directions under water, forgetting that the snorkel was not a megaphone, and nearly drowned in consequence. All in all, it was a most enjoyable and successful shoot.

Our next stop was to be Barro Colorado, but as we knew it would take the crew some time to get organized Lee and I decided to stay on for a few days in the San Blas Islands, since it is not often that you find such an ideal, unspoilt spot. However, I felt it incumbent upon me to go to Israel, our hotel owner, and remonstrate with him. It is not often that I do battle with a hotel manager, but in this instance I felt justified. After all, we didn't mind the sand on the bedroom floor, the fact that we had to make our own beds, if we could find the sheets, or the fact that the sea-water in the shower would suddenly cease owing to a surfeit of shrimps in the pipes, or that the lavatory (because two screws were missing) bucked like a rodeo horse and nearly precipitated you through the bamboo walls into the sea. No, we put up with these minor irritations because of the charm of the place. What we were really complaining about was the food. Breakfast consisted of coffee, toast, marmalade and cereals – perfectly satisfactory – but it was the other meals that filled us with despair. So, determined to be firm but fair, I talked to Israel.

'Israel,' I said, smiling warmly, 'I want to talk to you about the food.'

'Huh?' said Israel. One had to go fairly carefully with him, because his knowledge and command of English were rudimentary, so any sudden new idea inserted into his life was liable to panic him and make him as incomprehensible as Alastair.

'The food,' I said. 'Breakfast is very good.'

He beamed. 'Breakfast good, huh?'

'Very good. But we've been here two weeks, Israel, you understand? Two weeks.'

'Yes, two weeks,' he nodded.

'And what do we have for lunch and dinner every day?' I asked.

He thought for a moment. 'Lobster,' he said.

'Exactly,' I said. 'Lobster, every day. Lobster for lunch, lobster for dinner.'

'You like lobster,' he pointed out aggrievedly.

'I used to like lobster,' I corrected him. 'Now we would like something else.'

'You want something else?' he asked, to make sure.

'Yes, how about some octopus?'

'You want octopus?'

'Yes.'

'OK. I give you octopus,' he said, shrugging – and octopus he gave us, twice a day for the next five days.

The day we left, Israel suddenly appeared while we were sipping a farewell drink under the palms. He spouted a stream of his brand of English at me, speaking very rapidly and seeming, for such a normally impassive man, extremely upset. He kept pointing at the canoe that had arrived containing several women and children, bright and colourful as a boatload of orchids, with whom he had been carrying on a lively altercation. I gathered that we were not responsible for his wrath and I persuaded him to slow down and eventually managed to grasp the salient points of his story.

The previous evening, an Indian had paddled over from the neighbouring island which lay some three-quarters of a mile away, in order to celebrate some good fortune or other. He drank deep and late and eventually, at about ten o'clock at night, paddled off rather unsteadily in the direction of home. By dawn, as there was no sign of him, his wife borrowed a canoe and, taking her mother and family, went in search of him. All they found was his empty canoe floating over the reefs. Now they had arrived at the hotel to tell Israel he was a murderer for selling the man drink, and that it was his responsibility to find the corpse. Quite simply, Israel now wanted to know if we would go and help him to look for it.

Most women would have fainted if asked to do this – not my wife.

'How exciting,' she said. 'Do let's. We've got time, haven't we?'

'Yes,' I said, 'it would be nice to have a final swim with a corpse.'

As we were preparing to leave, the latest guest at the hotel appeared and approached us. She was a voluptuous, well-rounded lady with glossy black hair, glossy brown body and large quantities of glossy white teeth. Her suntan lotion could be smelt a mile up-wind, and she jangled musically as she moved from the glittering goldmine of ornaments she wore. What she was doing in the primitive San Blas Islands I shall never know. She looked as though she would have been much more at home on the Côte d'Azur or Copacabana Beach. The white bikini she was wearing was so minuscule she might just as well have not worn one at all.

'Excuse, please,' she said, giving us the benefit of all her teeth. 'You are going out swimming?'

'Er . . . yes, in a way,' I said.

'Would you mind if I come, too?' she asked beguilingly.

'Not at all,' I said heartily, 'but I must tell you that we are going out to look for a corpse.'

'Yes,' she said, head on one side. 'You don't mind?'

'Well, not if you don't,' I said gallantly, and she entered the boat, almost asphyxiating us with a combination of Chanel No. 5 and Ambre Solaire, tinkling like a musical box.

Israel steered us out to a new bit of reef, unfamiliar to us, where the canoe had been found. The bereaved family were already there, cruising up and down, peering hopefully into the water, which was some ten or twelve feet deep and glass-clear. Israel said that he would take one end of the reef and Lee and I should take the other. Miss Copacabana had already lowered herself elegantly into the sea and was hanging on to the side of the boat, looking singularly out of place.

'Will you help Israel, or will you come with us?' I asked.

'I swim with you,' she said, giving me a smouldering look. So the three of us set out. After ten minutes, we all met over a forest of staghorn. Lee had seen nothing, neither had I. Treading water, I turned to Copacabana Lady.

'Did you see it?' I questioned.

'See what?' she asked.

'The corpse,' I said.

'The . . . what?'

'The corpse. You know, the dead body.'

'Dead body?' she squeaked. 'What dead body?'

'That's what we're looking for,' I said exasperatedly. 'I told you.'

'Oh, Madre de Dios! A dead body? Here on the reef?'

'Yes.'

'And you are letting me swim with corpses?' she said indignantly. 'You are letting me swim with dead cadavers?'

'You wanted to come,' I pointed out.

'I go,' she said.

She covered the distance to the boat in record time and hauled herself on board.

'Oh, well,' said Lee philosophically, 'she would only have had hysterics if we had found it.'

It was time for us to catch our plane. We hadn't found our corpse. We had alienated Copacabana Lady. However, thinking about it later, I am inclined to think that what we did was somewhat foolish. After all, what better bait to attract sharks to an area than a plump Indian corpse?

So we took our tiny plane, and as it flew over the many small, palm-shaggy islands, like a scattering of green beads, and the smouldering reef showing like strange watermarks beneath the glittering surface of the sea we vowed we would come back someday to bathe and swim in this enchanted spot and to try to expand Israel's gastronomic endeavours.

SHOOT FOUR

OUR TRIP BY BOAT to Barro Colorado Island in the Panama Canal took half an hour and gave us our first view of the forest we were to work in. The launch chugged its way through the sherry-brown water, past a solid wall of multi-coloured trees. The canopy was thick, interwoven as ancient knitting, a smouldering mass of greens, reds and browns, with here and there a feathery pale-green tree rising above the rest, its silver-white branches starred with scarlet and emerald epiphytes and tangled bunches of purple-pink orchids. At one point, a pair of toucans, their huge beaks gleaming banana-yellow, flew fatly and heavily across our path; and when the launch was forced inshore by the shoals we could see humming-birds, like handfuls of opals, flipping among the tiny blossoms on the trees. The sky was a deep, rich blue, although it was only early morning, and the sun was hot enough to send trickles of sweat down your back under your shirt. That lovely, rich fragrant smell of the forest enveloped us, the delicate scent of a million flowers, a thousand thousand mushrooms and fruit, the perfume from a quadrillion gently rotting leaves in the simmering, ever-changing, ever-dying, ever-growing cauldron of the forest.

Presently, the island came into view. Hills like isosceles triangles forested to their very tips, their reflections blurred and smudged in the brown waters like a pastel drawing. As the launch drew in at the jetty, a Morpho butterfly, the size of a swallow, came and pirouetted around us briefly, like a piece of animated sky, before flying off to illuminate some dark-green part of the forest. We unloaded our gear, and then faced the fact that what lay before us was an almost one-in-one climb to the summit up a flight of cement steps, which reminded me unpleasantly of some of the steeper, more backbreaking Aztec monuments that Lee and I had crawled up a few years previously in Mexico. Alongside it ran a monorail with a flat-topped train-like engine on it. On this we piled our luggage and gazed up at the distant houses almost hidden in the trees.

'Well,' I said grimly, 'I'll walk up this once, just to say I've done it, but after that it's the Orient Express for me.'

I have rarely regretted a decision more. By the time I was halfway up I was exhausted and drenched with sweat. By the time I reached the top I had just enough strength left to stagger to a chair and grasp the tankard of beer which Paula so thoughtfully had waiting. Needless to say, to my intense annoyance, Lee had attained the summit looking immaculate and not in the least out of breath.

Since they had arrived, the crew had been busy checking suitable film-sites and finding out the best areas for animal photography. Most of the animals on the island were used to clutches of earnest scientists bumbling about in the forest so one more entourage would not make much difference.

'We've got good stuff so far. Well, when I say "good",

maybe . . . cutting-room floor, but . . . er, looks some good stuff, yes, some of those . . . roaring things . . . monkeys, yes, howlers and lots of those huge trees covered with epithets,' reported Alastair.

'Epithets?' I asked, wondering if this was some new form of parasitic plant I had not heard of, now growing in Barro Colorado.

'Yes,' said Alastair, 'you know, spiky things, like orchids.'

'You don't mean epiphytes, do you?' I asked.

'Oh, yes, I knew it was something like that,' said Alastair with fine aplomb, 'and then there were some of those things with long . . . er . . . noses, funny names.'

'Tapirs?'

'No, long noses, whiffly, kind of funny,' said Alastair, annoyed at my obtuseness in the face of such a detailed biological description.

'Anteaters?'

'No, no, no, they walk on the ground.'

'So do anteaters,' I pointed out.

'They call them something like "cocas",' said Alastair.

I thought deeply. Communication with Alastair was always difficult enough, but when he either couldn't remember the name or else used the wrong one you began to feel you were trying to unravel the Dead Sea Scrolls with the aid of a Portuguese–Eskimo dictionary.

'You don't by any chance mean a coatimundi?' I asked, with a flash of inspiration.

'That's it, that's it,' said Alastair triumphantly. 'Nose long, whiffly, climbs trees.'

Presently, we went off on our first foray into the centre of the island to see all the sets that Alastair had picked out, and to try to get a glimpse of some of the animals. However many times you visit the tropics, I don't think you ever get over the thrill of once again penetrating the dim recesses among the giant trees. Coming from a sun-drenched clearing outside, your eyes have to accustom themselves to the gloom. The first impression is of coolness, the cool dampness of a butter-dish; but you realize that this is only relative, for you are still sweating. The next thing to excite you is the great wealth of plants and trees around. Everywhere you look there is a new species and, although the riot of undergrowth is stationary, you get the impression of great movement. The giant trees, a hundred feet high straddling on buttress roots (like the flying buttresses of a medieval cathedral), are lashed together with a web of creepers and lianas, so that they resemble the giant masts of so many wrecked and abandoned tall-masted schooners, their green sails in tatters and only the shrouds of the lianas keeping them upright.

In places, the forest floor appeared to be alive, a moving carpet of green. This hallucination was brought about by the streams of leaf-cutting ants hurrying back to their nests with their booty, a triangular piece of thumbnail-sized green leaf, slung over their shoulders. From the tree of their choice (which they were busily dissecting) to their nest may be several hundred yards, and so these columns of green wend their way over the dark forest floor, over logs and under bushes in a steady stream that on close inspection looks like a Lilliputian regatta, all the boats having green sails.

As we made our way deeper into the forest, we could hear ahead of us the deep, vibrant roar that signalled a troupe of black howler monkeys. It's an impressive sound, somewhere between a howl, a roar and a harsh gurgle, and it shakes and vibrates the forest in a prodigious fashion. Presently we found them, a small family group, black as jet, some slouching nonchalantly through the branches, others lolling back in patches of sunlight, stuffing leaves and buds into their mouths, others simply hanging by their superbly prehensile tails and contemplating their aerial garden. When they caught sight of us, they became very alert, glaring at us suspiciously, and when we moved off the path into the forest so that we were directly underneath them they grew agitated and belligerent and broke off twigs and leaves to throw down at us, and less desirable ammunition as well.

'I say, that's a bit much,' said Alastair, as a large piece of excreta crashed through the leaves a few feet from his head.

'Now, cool your jets, Alastair,' said Paula. 'They're only doing what everyone wants to do to a director.'

The monkeys above us, having found that the barrage of twigs and excreta had no effect, now burst into a gigantic chorus to persuade us that this was their territory. It was like standing in the deep end of an empty swimming-pool listening to the Red Army choir, each member singing a different song in Outer Mongolian.

'We've certainly made them lose their cool,' said Paula, raising her voice above the racket.

'We must certainly, you know . . . howling, yes . . . somewhere high . . . trees,' said Alastair.

'There's a tower,' said Paula, doing an instant translation. 'They were telling me that there's a tall tower in the forest that they used to use for studying the forest canopy.'

'Just the thing,' said Alastair.

'About a hundred and fifty feet high,' said Paula enthusiastically.

'How delicious,' I said. 'I shall enjoy watching Alastair go up it.'

'Oh, honey, I forgot you don't like heights,' said Paula. 'Never mind, we'll send the crew up, and you and Lee can stay on the ground.'

'What a lovely producer you are,' I said.

We moved on into the forest, stepping carefully over the columns of leaf-cutters. So numerous were they that you wondered why the whole forest was not defoliated. This leaf-gathering is really a form of gardening, for the ants carry the leaves to their vast underground homes (sometimes a quarter of an acre in extent) and here they rot the leaves down into a mulch on which they grow the fungi which is their food. In some way realizing that, if they defoliate all the trees in the immediate vicinity of the nest, they would soon starve, they cull the trees carefully and only gather a certain amount of leaves from each tree.

On our second day we came to a clearing in the forest that had been created by the death of one of the giant trees. Growing on a slope, torrential rains had undermined its roots' tenuous hold on the topsoil and a gust of wind had then torn it free, as easily as a dentist wrenches a tooth from a jaw. It showed clearly why the tropical forest is so fragile. The topsoil is only a thin layer, so thin that the trees have to grow these giant buttress roots in order to keep upright. These huge trees, in fact, are feeding on themselves, for the moment their leaves fall they decay and become the humus on which the trees feed. So rapid is this process that only a thin topsoil is able to form. So the felling of the forest – as is happening at a horrifying rate throughout the world – exposes this thin layer, which only lasts a short time as agricultural or grazing land. Then it disappears and leaves erosion in its place. However, a natural tree-fall such as the one we found is a boon to the forest. As the giant crashes to earth, it splinters and fells smaller trees in its line of fall and tears a rent in the thick forest canopy. The sun floods in and the shrubs, creepers and baby trees, who have all been struggling in the gloom of the forest floor, shoot upwards in it. Seeds, which have been lying dormant in the humus for perhaps many years, waiting patiently for such an event, now sprout and start to rocket upwards towards the blue sky before the gap is closed by other plants. Thus the death of one of these forest mammoths is a signal for new life and growth around its huge carcass.

On the slopes above the fallen tree we heard a series of squeaks, chatterings and rustlings in the trees. Leaving the path to investigate, we found a group of spider monkeys disporting themselves low down among the trees, feeding on some pink buds. They are aptly named, for with their long, furry, dark limbs and their long tails (so prehensile that they use them as skilfully and as casually as if they were another limb) they did look rather like some strange giant spiders spinning webs among the branches. Unlike the unfriendly reception we received from the howlers, the spider monkeys seemed captivated by us and swung on their wonderful tails closer and closer and lower and lower. One in particular seemed specially fascinated by Lee, for she had just started to eat an orange to quench her thirst. He swung himself down from branch to branch until he was within fifteen feet of her, peering at her with all the dedicated interest of an anthropologist watching the feeding habits of an aborigine. Lee broke off a small piece of orange and held it out to him and, to our astonishment, without hesitation he swung down, grabbed the fruit and stuffed it into his mouth. After that, they followed us through the trees gazing at us wistfully and only going their own way when it became apparent that there were no more oranges forthcoming.

Alastair had arranged with one of the hunters attached to the station that he would comb the forest for suitable subjects for us to film, and the next day he came in with the first specimen, one of my favourite animals, the two-toed sloth. They really are enchanting creatures, their small heads, their shaggy bodies, their round, slightly protuberant golden eyes, and their mouths set in a perpetual, dreamy, benevolent smile. Slow and gentle, they will suffer you to hang them wherever you like, as though they were an old coat, and only after half an hour or so of deep meditation will they move perhaps six feet and that in slow motion. Sloths are really fantastic creatures. They are so beautifully adapted for their strange, topsy-turvy life in the tree-tops and, because they spend most of their lives upside-down, and because their diet is highly indigestible leaves, their internal organs are unlike those of any other mammal. Their whole metabolism is as slow as their movements, as slow as bureaucracy. They may go for a week without urinating, for example.

The sloth's fur, of course, grows differently from that of other animals. In other mammals the hair grows from the backbone towards the ground, so the parting, so to speak, is on the backbone. In the sloth, it lies along the side of the belly and the rest of the hair grows towards the backbone, so when the sloth is upside-down the rain runs off the fall of the fur more easily. They have a very strange adaptation of their fur – thin layers of cells which lie diagonally across the hairs forming ridges in which two species of blue-green algae flourish. This gives the animal's fur a greenish tinge, which acts as a camouflage among the leaves, so the sloth is, in effect, a sort of hanging garden.

Even more curious than this, there are several species of beetle and mite which have taken up residence in the sloth's fur, as well as a strange species of moth called the snout moth. There are approximately twelve thousand different species of this sort of moth scattered around the world and many of them are very curious. For example, some have what is called a tympanal organ on the base of the abdomen. This hearing organ can detect the ultrasonic cries of bats (developed to capture prey) and thus allow the moths to escape this predator. Some of the snout moths' caterpillars live on or in aquatic plants and in many cases become really aquatic, one species of caterpillar even developing gills. The species has a curious relationship with the sloth. It lays its eggs on the sloth's fur and when these hatch out the larvae feed on the algae which exist in the grooves and possibly on the fur as well, so as well as being a sort of hanging garden the sloth is also a sort of perambulating furry hotel for all these insects.

The next film star that was brought from the forest to

appear in front of the cameras was a fascinating little creature that I have not seen since I obtained some in Guyana many years ago. It was a pygmy anteater, the smallest of the anteaters, a beast that, fully grown, would fit comfortably into your cupped hand with room to spare. Like the sloth, this diminutive creature is perfectly adapted to its arboreal life. Its fur is short, dense and silky, an amber brown in colour. Its prehensile tail is naked at the end, which enables the creature to get a firmer grip with it, when it is wound round a branch. It has a short, tube-like snout, slightly curved, and tiny eyes and ears hidden in its thick fur. It is the feet of this little beast which are so extraordinary. Its hands are fat, pink pads armed with three long, slender, sharp claws, the middle one being the biggest. These claws can fold back into the palm of the hand like the blade of a pocket-knife. On the hind feet (the heel of the foot, so to speak) is a muscular pad shaped like a cup, which enables it to fit snugly round a branch. The toes on these feet end in sharp claws and have pads at their base so these, together with the suction-cup effect, help to form a prodigious clasping mechanism, without actually involving the claws. When in danger, the pygmy anteater lashes its tail round a branch, attaches its hind feet firmly (thus forming a triangle with the two feet and the tail), raises its arms above its head and, when its adversary is within range, falls forward, slashing downwards with its razor-sharp front claws. Unlike the sloth, who has blunt, peg-like teeth with no enamel that go on growing throughout its life, this anteater has no teeth, merely a long, sticky tongue and a very muscular gizzard in its stomach, which pulverizes the tree ants on which it lives.

Our specimen behaved with great fortitude during the filming and soon became so inured to us that, between takes, he sat quietly clinging on to Lee's forefinger, his tail carefully wound round her thumb or wrist. When the time came to release him, he was reluctant to leave Lee's hand and sat for a long time in the bushes, peering at us pensively, before moving away into the forest.

Although we had miles of film of the leaf-cutting ants going about their business of defoliating the forest, carrying their leaves back to their nest, cleaning out the nest and creating huge garbage-heaps, we had to part company with them when they vanished underground. This irked Alastair.

'I want . . . you know . . . I think . . . well, gardens,' he said, with his head on one side, revolving slowly, looking like a beaming, benevolent corpse on a gibbet. 'Mushroom-beds, you know . . . underground?'

'The only way you'll get them, honey, is by digging the guys out,' said Paula practically.

'Yes,' said Alastair musingly, moving top-like on the nest he was standing on, which covered an area approximately the size of a small ballroom.

'Is possible?' asked Roger. 'Is not too deep?'

'Well, sometimes the mushroom-beds lie quite close to the surface,' I said, 'but the ants won't take too kindly to it.'

'Paula, you get some spades and we dig, eh?' said Roger enthusiastically. 'Dig out ze little *jardins des champignons*, yes?'

'Yes . . . spades,' said Alastair, struck by the novel idea. 'Get some spades.'

So Paula traipsed back through the forest to the research station and eventually reappeared with a bundle of spades. The word 'producer' means exactly that. They are expected to produce – at the drop of a hat – anything from a four-wheel-drive truck to a square meal, a motor-launch to a bottle of whisky.

'Just the job,' said Alastair.

He and Roger seized spades and started to dig. Having had some experience of leaf-cutting ants, I took Lee and Paula by the arm and led them away from the scene of operations. Leaf-cutters, as a species, are highly successful creatures. The whole colony is founded by the queen, who, on her nuptial flight, carries (in a sort of pouch) a cluster of fungus threads which constitute the food for the future colony, in much the same way that the American pioneers used to take sacks of grain to plant when they eventually settled. When the wedding flight is over, the queen plants the fungus in a brood chamber and looks after it with all the dedication of a horticulturist, manuring it with her excrement. If the fungus dies, the colony fails; when it is successful, the colony expands and grows in proportion to the fungus gardens and may eventually have more than a million individuals to a nest. I had just explained this to Paula when approximately half the million inhabitants of this nest decided that the activities of Roger and Alastair were inimical to their well-being, so they poured forth to remonstrate. One minute Alastair and Roger looked like two earnest gardeners turning over their asparagus-beds in preparation for a new crop and the next minute they were executing leaps and twists and *pas de deux* that would have been the envy of the Moscow Ballet. This was accompanied by wild, tremulous screams of agony, interspersed in equal parts with blasphemy and procreative oaths.

'Christ,' shrieked Alastair, waltzing around, now of necessity. 'Ouch, ouch, they're biting. Oh, the bloody things!'

'Ouch, ouch, merde alors!' screamed Roger, waltzing, too, and slapping his trousers. 'Zey is biting.'

The chief problem was that Alastair was wearing shorts and an ancient pair of baseball boots, and this did not give his legs any protection, so the ants swarmed up him as though he were a tree, attempting to tear him to pieces. Roger, if anything, was in worse case, for he was wearing elegant, fairly tight-fitting trousers, up which the ants flowed with speed and precision. Those on the outside bit right through the thin cloth and into his flesh. Those on the inside concentrated on getting as high as possible before beginning their assault, so that Roger was being bitten in the most intimate and tender parts of his anatomy. The ants' jaws, powerful enough to chop up tough leaves, made short work of the thin trouser material and Roger's legs were patched with bloodstains as were Alastair's legs. We got them both away from the immediate scene of battle and de-anted them. Paula then practised first aid with antibiotics, but it was a considerable time before we got all of the ants off them.

'Did you see that?' panted Alastair, his spectacles misted over with emotion. 'The buggers were trying to defoliate me.'

'What about me?' said Roger. 'Me they go for the private parts. Me they try to make eunuch.'

Later on, wrapped in so many layers of clothing that they looked like Tweedledum and Tweedledee clad for battle, they succeeded in unearthing a small section of the mushroom-garden and filming it, to the ants' fury.

One of the most fantastic pieces of natural history in the forest, one that was in its own way just as difficult to obtain on film as the ants' fungus-garden, was the extraordinary story of the giant fig tree and the minute fig wasp. This strange relationship only recently became unravelled, and it shows part of the enormous complexity of the tropical forest and how any plant or creature is only part of the whole intricate ecosystem, for without the great fig trees the fig wasp would perish and without the fig wasp the fig tree would never reproduce its kind, and its numbers would dwindle and it would eventually become extinct.

All figs have a very curious flower structure, resembling, in fact, that of a fruit more than that of a flower. A host of tiny flowers lie inside the fig, which is attached to the tree by a stalk at one end; at the other there is a minute opening almost obscured by scales. Figs have male and female flowers and the way the pollen is ferried from one to the other is as enchanting as it is awe-inspiring. This is what happens:

In the fig the first to mature are the female flowers, and their scent attracts female fig wasps, who are carrying pollen from other fig trees in the forest. To get at the blooms, the wasp must climb inside the fig using the opening at one end, shouldering the scale 'door' open. This is not an easy process for the door is stiff, and the female wasp is fragile and often loses her wings and antennae when entering the fig.

Once she (and other females) has successfully broached the fig, she proceeds to bore down through the styles of the female flowers using her long ovipositor, like someone drilling for oil. The flowers are of two sorts, one with short styles and one with long styles. This design is such that the ovipositor of the wasp can only reach the ovules of the short styled variety to lay the eggs. The long styled ones are only probed, but while being probed they receive the pollen carried by the wasp. So, by this process, the short styled fig flowers produce fig wasp larvae, whilst the long styled flowers produce seed. This is extraordinary enough, but the story gets even more bizarre and magical.

The next thing is that the larval wasps develop and then pupate. At this stage they apparently produce a substance that prevents the fig from maturing, for if it were to ripen while the wasps were pupating, their nursery might be eaten with them inside. At last the pupae mature. The males are the first to hatch; they go the rounds of the as yet unhatched females to fertilize them. Up to this point, to all intents and purposes, the fig is completely sealed so that the atmosphere inside it contains up to ten per cent carbon dioxide (as opposed to .03 per cent outside it), but this does not seem to worry the males. However, after mating the male wasps tunnel through the sides of their nursery and the carbon dioxide level falls dramatically. This in some way accelerates both the hatching of the female wasps and the emergence of male flowers, and the females get coated with pollen from them. Both the male and female wasps, working as a team, bite away the scales at the end of the fig, and the females fly off, carrying fertilizing pollen and stored sperm to found a new colony in another fig tree at the female stage. The males, being wingless, cannot leave the fig and so they die, their life's work done.

When you think that the fig wasp story is only one of the many fantastic things that are being discovered in the tropical forests everywhere, it makes you realize what a complex world we live in and how our ham-handed tinkering can cause havoc with the delicate balance of the ecosystems.

The tropical forests of the world are one of man's greatest bounties and yet the way we are treating them you would think that they are dangerous to us, instead of being an enormous self-generating storehouse of medicines, foodstuffs, timbers, dyes, spices and a host of other things. We do not as yet know the full benefit of the tropical forests to mankind, yet we are destroying the forests so fast that species of plant and animal are becoming extinct even before they are scientifically described. It is estimated that this wild suicidal attack on the tropical forests of the world is resulting in 110,000 square kilometres – 43,000 square miles – of trees being felled and burnt each year. Cheerful prediction is that at this rate all this kind of forest will have disappeared in eighty-five years. If this happens – and there is no indication that mankind is suddenly going to give up stupidity and behave sensibly – the alteration to the climate may well be catastrophic, for forests control the weather and without them you can turn rich areas into deserts in a very short time. This is to say nothing of the benefits to us that we have already discovered in the forests and the ones that await discovery. We have only touched the fringe of knowledge when it comes to that enormous ecosystem known as the rainforest or jungle. What inestimable benefit for mankind may lurk among the trees we have no idea, and yet in a profligacy that is almost maniacal we are destroying something that can never be re-created, something of tremendous value to mankind; something, moreover, which is self-renewing if it is husbanded and exploited with care. However, at the rate we are going, it is probable that in less than a hundred years with millions more mouths to feed we will be faced with deserts on which to grow food, simply because we are behaving in a greedy, malicious and totally selfish way, and this goes for everyone, regardless of colour, creed or political persuasion, for unless we move and move fast our children will never have the chance to see that most fascinating and important biological region of the planet, the tropical forest, or to benefit from it.

SHOOT FIVE

IT WAS PROBABLY just as well that we did not go directly from Barro Colorado (where the temperatures were in the hundreds) to our next location, for the temperature was thirty degrees below zero when we arrived at Riding Mountain to film the boreal forest in winter. To say it was cold means nothing. On leaving the car, my beard and moustache froze together and Paula's half-inch long eyelashes had such a thick coating of ice on them that she had great difficulty in keeping her eyes open.

'Jee-sus,' she said, surveying the snowbound countryside, the ice-grey sky and the enormous log cabin we had arrived at. 'Some place to hang up your shingle – huh, guys?'

'Um . . . snow, I hope . . . though the light's not very good. Will the snow affect the animals?' asked Alastair, trying to walk in a circle with his head on one side and being defeated by the depth of snow. Lee threw a snowball at him and, unfortunately, missed.

The log cabin belonged to Bob and Louise Sopuck, a charming couple who were to be our landlords and general helpers on this leg of the trip, together with their neighbours, Cheryl and Don Macdonald. When I say it was a log cabin, I do not mean one of those shoebox-shaped edifices you so frequently see in films about the old West. This large building was certainly constructed out of giant pine logs, but there the resemblance ended. Once you entered the cabin you saw that the whole of the lower floor was open-plan. The roof soared thirty feet or more above you and below were the dining, living and kitchen areas. Stairs led up to bedrooms under the eaves, and more stairs down to more bedrooms and cellars below the house. It was one of the most unusual houses that I have ever been in. Bob and Louise were both excellent cooks and had prepared a sumptuous meal to thaw out our bodies, frozen after our four-hour drive from Winnipeg: rich soup, home-made bread, followed by so much venison that I wondered if the entire deer population of Canada had been eliminated in our honour. Actually, Bob manages their land excellently and every year only shoots just enough venison to last them through the winter (generally two animals), and this acts as a beneficial cull as well as supplying delicious food.

With the aid of the food, accompanied by a certain amount of medicinal Scotch which we had thoughtfully brought with us, my beard and moustache thawed out and Paula was able to see once again. It was then suggested that she, Rodney (our cameraman in Canada) and Alastair went off on a recce, while Lee and I were introduced to the mysteries of skiing and snowshoeing. I very soon discovered that my figure, my *en bon point* as the French so beautifully put it, was not designed for skiing. If I leant forwards, I fell flat on my face; if I leant backwards I fell flat on my back; if I maintained a regal upright stance, I fell either forwards or backwards, depending on which way the wind was blowing.

Snowshoeing, however, was quite another kettle of fish and most satisfactory. It was a fascinating feeling, to be able to walk over the deep snow without sinking in. It made you feel like one of those lovely birds, the lilytrotters (albeit a fairly massive one), who progress over the waterlilies as though they were a highway. With shoes like tennis rackets you could progress at a steady pace over a depth of snow that, without the shoes, would have had you floundering and immobilized within a few feet. Turning round on shoes was the only tricky part, and if you did not do it right your shoes got tangled up and you fell into the snow, from whence it required much struggling and effort to get yourself upright again. Once you had mastered it you felt like a guardsman with outsize feet, doing a smart right-about-turn on the parade-ground.

Having acquired some expertise on the snowshoes, Lee and I went off to explore the immediate environs. The sky was slate grey, so solid it looked as if it would clang like iron if you threw a snowball at it. Out of it drifted handfuls of snowflakes, each the size of a penny stamp, as thick and as soft as blotting-paper. The snow squeaked and purred under your shoes, but apart from this the silence was complete, the world gagged with snow. The pines looked as though some giant pastrycook had flung icing sugar at them, their dark-green branches bending under the weight. In places you could see where quite large trees had been bent over by their white load and it was obvious that with the next snowfall the trees would be wrenched from the ground. We came to a small lake, round and smooth as a saucer of milk under its ice and snow covering. At the edges we could see snow-covered hummocks with black branches sticking out of them, like sticks of charcoal breaking the icy crust. These were beaver lodges and deep inside the animals slumbered, waiting for the spring to melt the five-foot piecrust of ice and snow and thus release the water for them to swim in.

In the summer, when we returned to Canada, we revisited this lake at dawn. Then the change was spectacular. The water was greeny-gold and the whole lake was rimmed with thick beds of reeds, like a fringe on a Victorian tablecloth, and here and there the surface was embroidered with patches of white waterlilies. The sun had just lifted above the green, shimmering trees, pulling wisps of mist up from the lake's surface, delicate skeins drifting among the reeds and the waterlilies, like fragile wedding veils.

We took a canoe, and travelled slowly out across the water towards the brown hump, like a giant, badly made Christmas pudding, that was the beavers' lodge. Halfway to it, a large brown head suddenly broke the surface of the greeny-gold water, and in a circular picture-frame of ripples a beaver contemplated us with a certain suspicion. We paused in our paddling to watch him as he swam slowly and sedately to and fro in front of the lodge, like a guardsman patrolling in front of a palace. When we attempted to manoeuvre the canoe closer to him, however, he panicked and lifted his paddle-shaped tail out of the water and brought it down on the surface with a blow that echoed across the lake like a gunshot and then dived. A few minutes later, he appeared in a different place and, seeing that we had not retreated, he smote the waters again before diving once more. The whole time we were out in the canoe, he kept reappearing, each time in a different place, and smiting the water to frighten us off. He was the only beaver we saw during our time in Canada and I cannot say that he behaved in anything like a welcoming manner.

Back at the house, we found Alastair in high spirits, for he had come upon and filmed a large herd of white-tailed deer and a somewhat recalcitrant moose which proved, as far as he was concerned, that there were animals in this frozen wilderness. As we had only seen two crows in our four-hour drive from Winnipeg, I must say I hardly blamed

Alastair for his belief that the frozen north was bereft of all life other than human.

'Tomorrow, we will go out and try to get some shots of you and Lee with the white-tailed deer and moose,' said Paula, producing. She had forgotten the days in Madagascar when she was merely the assistant producer, and went under the nickname of 'Ass. Prod.'

'Then in the evening,' she went on, 'we will go down to the lake where Alastair wants you to fish for owls.'

'I beg your parden?' I said.

'Fish for owls – with a mouse,' explained Paula.

'Quiggers, how much have you had to drink?' I asked.

'No, no, honey, I'm serious. Alastair has read somewhere that scientists fish for owls with dead mice as bait, in order to catch them and ring them or something,' said Paula.

'Never heard such rubbish,' I said, 'and, anyway, why at the lake? I didn't know that Canadian owls were aquatic.'

'No, it's just that there's more space on the lake. In the woods you might get your fishing-line tangled up in the trees.'

'I don't know. It seems mad to me,' I said. 'Can't you control Alastair?'

'No,' said Paula simply.

That night, Lee and I had our first experience of the Northern Lights. Because they were so commonplace, as far as Bob and Louise were concerned, they had not thought even to mention them. They had kindly installed us in their own bedroom, and when we got into the large, cosy double bed we found that directly above us was a huge skylight. I switched off the light and was immediately transfixed with astonishment. The large area of sky immediately above our bed appeared to be alive. Against the deep soft blackness of the sky were etched scrolls, curtains, scarves and tangled wisps of pale purple, green, blue, pink and frost-white fronds of what looked like cloud but which seemed to have a life of their own. With each passing second they shifted, separated, merged, broke up and re-formed in a different pattern and always they were floodlit from somewhere in the wings, as it were, and the colours changed with their movements. I was reminded irresistibly of a kaleidoscope that I had been given when I was a child, a triangular tube like a microscope. Beneath the lens you put patterned paper, particularly the garish, glittering paper from chocolates, and as you moved the tube about beneath the lens patterns shifted in a miraculous way. Now it seemed to me that the skylight was like the eyepiece of my old kaleidoscope and without any effort on my part was producing these fantastic effects in the sky, effects far more subtle and miraculous than any that chocolate-paper could have produced. We lay and watched this incredible display for an hour or so, until finally it dwindled and died, just leaving the velvety moleskin sky freckled with stars. It was a good thing it had died away, or we would have watched it until dawn and then been too tired to get up. It was one of the most eerie, delicate and beautiful phenomena I have ever seen.

Early next morning, after a gargantuan breakfast, we set off into the forest, muffled up in so many clothes that we felt as awkward moving as moonwalkers without the aid of low gravity. The grey weather that had greeted us had vanished in the night and the sky was flax blue and the faint heat of the sun made great mushroom-tops of snow slide off branches and land with a soft sigh on the carpet of snow beneath the trees.

One of the things we wanted to try to film was the extraordinary life that goes on beneath the smooth thick layers of snow. Not long ago it was discovered that the first layer of snow to fall – when it is covered by other layers – changes composition. The flakes seem to fuse together and produce what are to all intents and purposes icicles or ice crystals, so this layer of snow becomes transformed into tunnels and ice palaces a-glitter with delicate crystals. This is called the pukak. These ice corridors and palaces are, of course, a few degrees warmer than the outside air, since the overcoat of snow acts as an insulation. So the mice and other small rodents can live comfortably in the pukak, digging down to find grass roots for food, and some delicate things like insects or spiders can pass a comfortable winter in these corridors in semi-hibernation. In order to show the wonderful insulating properties of the snow, Alastair wanted to build what the North American Indians call a quinzhee, or what the Eskimos would call an igloo. So, while we and half our entourage went in search of animals, the other half, led by Alastair, found a suitable spot and started to build a quinzhee.

We had not driven far down the road when we spotted a moose with monstrous chocolate-coloured horns, standing among the trees at the side of the road. Moose are such curious-looking creatures with their ungainly bodies and legs and their bulbous wine-bibbers' noses. They always look to me as though they have been made up of discarded bits from several different creatures. This animal gazed at us lugubriously for a few minutes, twisting his ears; and then, blowing out two great clouds of steam from his balloon-like nose, he moved off heavily through the trees. The adult male moose is of course a magnificent animal, as large as a shire horse and with enormous palmated antlers, like giant holly leaves, on his head. We watched them later on in the spring, grazing along the lakes and rivers on the waterlily roots. Their heads and huge antlers would be plunged beneath the surface of the water as they grazed and then when they came up for air their antlers would be entangled and decorated with waterlilies and their stems.

We drove on and then ten minutes later two huge male elk came into view standing majestically by the side of the road, their branched antlers like some magnificent bony candelabra on their heads. They gazed at us with regal disdain for a moment or so and then trotted off slowly and gracefully, threading their way through the trees adroitly so that their massive antlers did not get entangled in the branches. Just as they disappeared, a whole herd of white-tailed deer, gingerbread-brown, came trotting into view, ears pricked, nostrils wide, big liquid eyes peering fearfully across the snow. They came to a halt when they saw us, clustering together nervously, their noses testing the air. For a moment or so it seemed as though they might actually cross the road in front of us, but then one more nervous than the rest panicked and in a moment the whole herd had turned in a flurry of snow and were off, their pale backsides like strange heart-shaped targets bobbing among the charcoal-black trees.

We drove on through the glittering frozen landscape under the bright-blue sky and half an hour later we saw, ploughing its way across a white valley between two black belts of leafless trees, what appeared to be a chestnut-brown avalanche. As we drove nearer, it proved to be, to our excitement, a small herd of six buffalo, hunchbacked, shaggy, wading shoulder high through the snow, packed tightly together, trailing shawls of white steamy breath behind them. As they ploughed through the crisp, white, undisturbed valley, leaving a broken path of churned-up snow and blue shadows behind them, they did look very like an avalanche of curly fur, muscular shoulders and glimmering horns.

We watched them for perhaps ten minutes until they were out of sight. We were just going to start up the car when suddenly, from the interior of the dark mesh of trees, out strolled an enormous old bull buffalo. On to the snowfield, white as a banqueting-cloth, he sauntered out, his beard swinging to his rolling walk, his horns sharp-curved as bows, his great forehead and massive shoulders a mass of dark ringlets, the breath from his nostrils making two cumulus clouds of steam ahead of him as he moved. Slowly, like a portly, well-made man of substance taking his constitutional, he moved across the white expanse. Here the snow was not so deep, so it only came up to his knees. He moved ponderously across until he was perhaps two hundred yards from the tree line. Then he paused and mused, his breath forming a cloud around his face, entangling itself in the fur of his forehead and shoulders. Then, in slow motion, as it were, he doubled his feet under him, and lay down in the snow. He lay there for a moment, and then with a vigorous kicking of his legs he rolled over. More kicks rolled him back again and so for the next ten minutes we were privileged to watch him take his snow bath – rolling to and fro, grunting with the exertion, his breath spouting silver clouds into the air, the snow flying in white cakes in all directions. Presently, exhausted by his ablutions, he lay on his side for a time, panting, his flanks heaving. Then he hauled his bulk upright, gave a gigantic shuddering shake that made the snow burst from his thick fur in a cloud and then, massively sure of himself, he sauntered after the herd, of which we had no doubt he was king. Slowly, meditatively, like a huge, dark cloud, he moved across the snow and disappeared.

We got back to find Alastair, pink with exertion, proudly walking round a five-foot conical mound of snow.

'Quinzhee,' he explained proudly, with his head on one side and contemplating the pile of snow with affection. 'Got to . . . you know . . . snowshoes and dig it.'

So with our snowshoes we patted the snow flat all over the pile and then proceeded to dig a large opening like a small church door on one side. Through this, we dug deeper and deeper until we had hollowed out the whole of the interior. It was interesting to see how the snow on the outside of the quinzhee was just snow, as it were, whereas the layers underneath which we had dug through were already turning crystalline and forming the insulation. Crawling into the quinzhee, Lee found that, although the temperature outside was minus thirty, inside our snowhouse it was one degree above freezing – not a great deal in terms of cold, perhaps, but sufficient to save your life if you were marooned overnight in this harsh environment.

We had just finished the quinzhee when the various birds arrived to see what we were up to. The first was a group of chickadees, fragile, tit-like birds, so delicate you wondered how they survived the rigours of the winter. They played about in the trees, hanging upside down and chirruping at us, but eventually got bored and flew off. The next to arrive were a group of evening grossbeaks, beautiful, heavy-beaked finches clad in startling gold and greenish-black plumage, flashing in and out of the dark pine branches like little golden lights. They seemed much more nervous than the chickadees and soon flew off into the sombre depths of the forest. Our next bird visitor by contrast was much bolder. It was a whiskyjack, a medium-sized jay, handsomely clad in pale greys and blacks. He suddenly appeared flying out of the forest and alighted on a tree nearby. He hopped from branch to branch, pausing now and then to watch us with his head on one side, reminding us irresistibly of Alastair. Whiskyjacks associate humans with food and therefore are the boldest of the forest birds. A search among all our pockets brought to light the remains of some biscuits and a handful of peanuts. These were held out to the jay, and to our delight he flew down quite confidently and perched on our fingers, stuffing as many morsels into his beak as possible. When he had a beakful he would fly off, and what he did then was extraordinary. Finding a suitable branch, he would stick the food to it using as adhesive his ultra-sticky saliva. In this way he gathered all the peanuts and biscuits we offered him and made perhaps seven or eight caches of the food in various trees – larders for the future. He seemed a trifle annoyed when we eventually ran out of foodstuffs, but by that time we reckoned he had stashed away enough food to keep ten jays going for a week.

It was getting towards evening when we got back, and Alastair was anxious to film the owl-fishing sequence. Bob had provided a fly-fishing rod and as bait two stuffed mice. Solemnly, we all trooped down to the frozen lake and made our way out on to the ice. Here Bob gave me a swift lesson in casting, since I had never used a fly-rod before. He showed me the wrist action required several times, making the mouse land on the ice some thirty feet away as lightly as a feather. It seemed perfectly simple to me, and I could not see why fly-fishermen made so much fuss over casting. I seized the rod with confidence, pointed the tip skywards and made what I considered to be a perfect cast.

Unfortunately, for some reason, the wretched line, instead of unfurling and gently lowering the mouse on to the ice, behaved like a whip, with the result that the mouse snapped in two and half of its body went soaring away across the pond, leaving me with only the head and forelegs still attached to the line.

'Honey,' said Paula, when she had stopped laughing, 'you know the budget can't afford endless mice.'

'Lucky we've got a spare,' said Alastair.

'I think I had better just practise with this half for a bit,' I said. 'I don't want to ruin the other one.'

'I think, when they see this bit of film, they will say that the mouse was miscast,' said Alastair, and went into convulsions of mirth at his own wit.

Austerely, I ignored his vulgarity and took myself over to a quiet corner of the pond to practise with my demi-mouse. Once I had mastered the art of casting, we attached the whole mouse to the line and filmed the sequence. Needless to say, not a single owl came anywhere near us.

That night, our last in Riding Mountain, we were again treated to a wonderful display of the Northern Lights, and for two or three hours we lay in bed watching the sky above us form ribbons or scrolls or fluted curtains that glowed in pastel shades as though there was a fire within them, merging, parting, disappearing, reappearing, an apparently never-ending and never-repeated pageant. It was quite extraordinarily beautiful, and one longed to be able to paint it, although you knew that a painting could never convey the magical elegant patterns etched in the sky. For the Northern Lights alone, I felt it was worth braving the Canadian winter.

Our next visit to Canada was in the summer, when the countryside presented a very different mien, with trees in full leaf and flowers growing everywhere. Our destination was Banff, one of Canada's major national parks, situated in the middle of the Rocky Mountains and so containing some of the most spectacular scenery in the world. Mountain range after mountain range, like a giant stormy sea sculpted in rock. Pine forests like green fur crawling up their flanks and snow glinting on their peaks, with here and there, as though a candle had melted wax down a cliff-face, a baby glacier clinging immobile. The park was wonderful, for each time you rounded a corner a spectacular and lovely vista of mountains regaled your eye and you thought that that must be the most beautiful view of the park, only to be proved wrong when you rounded the next corner and were presented with something more stupendous. Here and there, the mountains rose so sheer that only their feet were tree-covered, but each brown or cinnamon-coloured peak had snow carefully arranged on its valleys and along its ledges like crisp, freshly laundered napkins laid out on the desolate rocks.

We paused at a layby for a rest, and found beneath the trees a profusion of tiny wild strawberries, glowing like little lanterns in the darkness of the leaves, on which we gorged ourselves. Apparently the grizzly and black bear, both of which inhabit the park, are inordinately fond of these strawberries and, if you are deep in the woods, it is wise to keep an eye out in case you suddenly find yourself sharing a strawberry-patch with a grizzly.

Above us rose two great carunculated axeheads of mountains, and between them lay a valley shaped like the bowl of a spoon, jade-green with vegetation. On this green surface were some white specks, which I at first thought were patches of snow, and then I saw one move and realized it was a small group of an animal I had long wanted to meet – the Rocky Mountain goat. Don't be misled by the name. The Rocky Mountain goat is the king of goats; with his soft white coat, softer even than cashmere, and his black hoofs, horns, muzzle and eyes, he is a dandy of a beast. Unlike the other mountain-dwelling ungulates, he is not hurried or panicky in his movements, but sedate and slow and surefooted. So surefooted, in fact, that one was seen to attempt a leap from one narrow shelf on a high cliff to another, which proved to be just too far away. Instead of crashing into the gulf as any other animal would have done, the goat, realizing its mistake, changed its attitude in mid-air, hit the rock-face with all four feet, did a backward somersault and landed safely on the shelf it had just vacated. They have few enemies, and those that they have they seem well able to deal with. One Rocky Mountain goat, beset by hunting dogs, killed two with its horns, pushed another over the cliff to its death, and when the rest of the pack became faint-hearted and retreated the goat sauntered off as if nothing untoward had happened. On another occasion, one was found dead, killed by a grizzly bear. However, near the goat was found the corpse of the grizzly, stabbed twice neatly through the ribs close to the heart. He had obviously just had the stamina to kill the goat before dying of the mortal wounds inflicted on him by the dagger-like horns. With our binoculars, we watched them for some time as they grazed across the brilliant green grass but we saw no exciting incident involving a grizzly. They grazed peacefully, occasionally pausing to stare around them, their long, earnest, pale faces giving them an air of sobriety and respectability, like a flock of vicars in white fur coats.

One of the things we were anxious to film was the summer activities of the pikas, strange small rodents that live high up in the alpine meadows. These little animals do not hibernate during the winter months as so many of the mountain creatures do – like the fat marmots, for example. Instead they have become farmers, and during the summer months they feverishly collect grass and leaves which they pile into haystacks that dry in the sun. When one side is sufficiently dry, the haystack is carefully turned by the pika, so that all the collected food gets its share of sunshine. These haystacks are put in sheltered places and during the winter months form the larders of the pikas, without which they would starve to death when the valleys are snowbound. At the first sign of rain, the haystack is moved under cover and put out in the sun again when the storm is over.

According to Geoff Holroyd, the young man who was acting as our guide to the region, the best place to watch pikas at their farming activities was in an alpine meadow some twenty miles away from the hotel in which we were staying. So early in the morning we set off. When we arrived at the base of the mountain range, we left the car and started up the two-mile, almost vertical climb through the alpine larch and pine woods. As we started to climb, we heard all around us what we took to be the whistling call of some bird which seemed to haunt the forest in profusion, judging by the number of calls we could hear. Presently, however, we came to a small clearing and there, squatting at the mouth of its burrow was the musician responsible for the ringing, flute-like cry – a fat ground squirrel, wearing a tasteful suit of rust-red and grey fur. He sat as upright as a guardsman at the entrance to his home and his ribcage pumped in and out as he gave his musical warning cry. His big liquid eyes stared at us with that intense, slightly inane expression that most squirrels wear, and his little paws trembled with his vocal efforts. He was not really all that afraid of us, for he allowed Lee to get within four or five feet of him before he retreated to the safety of his burrow. These were the Columbian ground squirrels and, as Geoff explained, you got different species of squirrel at different levels on the mountainside, so roughly speaking you could tell how high you were by the species of squirrel you were surrounded by.

As we climbed higher, the larch and pines started to thin out, and at the timber line, where the forest started to give place to small arctic plants, the trees had become stunted and dwarfed by the freezing winter winds and the savage pruning by knife-edged particles blown by it, turning pine and larch into pygmy trees, resembling the miniature Japanese bonsai. Here and there among these mini-forests were flashes of yellow, tomato pink and deep red where the alpine paintbrush grew, an elegant and beautiful plant, stem and leaves covered with a mist of fine hairs. These hirsute coverings can be found on a lot of alpine plants and, believe it or not, actually help to protect the plant from at least nine months of icy weather in the same way that the thick coat of the grizzly bear protects its body.

Here the trees ended and the valley stretched out before us, lush and so green it would have put a mine of emeralds to shame. On the flanks of the mountainside surrounding it you could see the scars of old avalanches, but the bright meadow itself seemed unscathed. Here the rich green grass was spangled with tiny alpine plants, the bright-yellow cinquefoil, the yellow heather, the delicate purple milk vetch, the white sandwort and cushions of bright-pink moss campions. Through the centre of the valley ran a stream chattering and glistening around piles of grey rocks, each so beautifully arranged that it seemed as if the whole valley had just been completed by some alpine Capability Brown.

Suddenly our attention was attracted by a shrill whistle that echoed back from the surrounding hills, and on top of a pile of rocks, lolling in the sun, was a fat, brown marmot, looking somewhat like a gigantic guinea-pig, with a long furry tail. I think he gave his alarm cry from force of habit, for he did not seem at all upset by our presence. In fact, after being near to him for a few minutes, he actually allowed me to approach him and scratch his fat neck and tickle his whiskers. It was a marvellous, magical feeling to be in a place where the animals look upon you as being benign and allow you to share, however briefly, a part of their lives.

Ahead of us, we could see the valley was divided into two – a higher and a lower area, separated by a steep slope of tumbled boulders. Beyond lay a barren hillside of jagged rocks, arranged in piles by past avalanches. These huge boulders were embossed with fossil seashells and coral – an indication that in times past these rocks had formed the bed of an ancient sea that had, by a cataclysmic movement of the earth, been lifted up to this high and remote valley. It was here, under a huge tottering fortress of fossil-decorated rocks, that we saw our first pika. It must have been sitting among the rocks and observing us for some time, for its ash-grey fur made it blend in with its surroundings so beautifully that it was quite invisible until it moved. About the size of a guinea-pig, it had a very rabbit-like face, but its eyes were larger and darker and its ears were small and shaped like half a coin. It had no discernible tail and its fur was glossy like satin. It stared at us, uttered a shrill, piping cry of disquiet, leapt down among the rocks and vanished. Investigating, we soon found several haystacks about two feet in diameter and about twelve inches high. Most of them had fresh grass or leaves on top, so it was obvious that the pikas were still in the process of harvesting. Curiously enough, in this lower valley the pikas were very nervous, only showing themselves in flashes when they scuttled from crevice to crevice. When we climbed to the upper valley, however, we found a most amenable and enchanting pika who was far too busy with his farming to take any notice of our presence. Here the stream had carved itself out a bed in the thick, spongy green turf starred with flowers, and it twisted and wriggled its way across the meadow like a plait of grass. In the middle of the turfs, smooth as a billiard-table, a fat little pika was sitting, busily tearing out mouthfuls of grass, which he kept in his mouth. Eventually, when his jaws could hold no more, he sprinted off to his house and haystacks among the rocks, looking incongruously as though he was wearing a large green walrus moustache. We followed him into the scree and found his home was under a gigantic rock the size of a small car, and nearby were two completed haystacks, and a third in the process of construction. He was so busy about his husbandry that I sat on a rock within three feet of his haystacks and he did not even pause in his serious work. As soon as he had arranged the mouthful of grass to his satisfaction with his paws, he was off through the rocks speedily, bouncing like a rubber ball, until he reached the meadow. Here he collected another mouthful of grass and made his way back to where I sat. Within a couple of feet of me he paused, peered at me with dark eyes over his monstrous green moustache, and then unconcernedly went to his haystack and added his burden to it, patting it carefully into place. The curious thing was that when the pikas in the lower valley uttered their shrill whistling alarm call our pika would stop and whistle back, even though, on several occasions, he was so close to me that I could have reached out and touched him. He treated the fact that there were seven human beings clustered around him filming and recording his every move or sound with complete sang-froid, ignoring us and getting on with his busy schedule as if his life depended on it – which, of course, it did. The only point at which he got a trifle agitated was when we produced the kites, which sent his brethren down the valley into whistling hysterics.

Somewhere in his researches, Alastair had come across an experiment Konrad Lorenz had conducted on some chickens. He had constructed a large cardboard silhouette of a flying goose. When he passed this across the sky above the chickens, they took no notice. However, if the silhouette was reversed, that is to say it was flying backwards, the chickens reacted strongly. The silhouette became that of a hawk, and the long neck and head of the goose became the hawk's long tail, and the tail of the goose was the small round head of the predator. Nothing loth, Alastair was determined to have us fly some kites to see what effect it would have on the pikas, and so he had paid a visit to a shop in Toronto's Chinatown and had procured two most elegant kites depicting hawks. Once we got these really beautiful kites airborne, the effect they had on the wildlife of the valley was considerable. Marmots shouted obscenities at them, pikas in the lower valley appeared to suffer a collective nervous breakdown, and four ptarmigans who had been walking along sedately minding their own business ran among the rocks and disappeared by the simple expedient of crouching down. Our pika, although he viewed the paper hawks with some alarm, was such a dedicated farmer that he did not let their presence interfere with his haymaking. He would pause now and then and utter a slightly muffled whistle through his green moustache. Once, when a kite's shadow slid across the green grass towards him, he did dive for cover but soon reappeared and continued his activities. I think of all the wonderful places I visited and enjoyed in Canada, our days spent flying kites in the Kamaskee alpine meadow were my favourite. The air like wine, brilliant sunshine stamping blue shadows on the mountains and the avalanche cicatrices. The vividness of the colours, the sparkling clarity of the little streams, and the quietness and peacefulness of the place were wonderful. It was one of those places you find in the world which are so magical that you wish you could live there.

SHOOT SIX

THE CANADIAN FORESTS ARE, of course, most spectacular, but the English woodlands can be extremely interesting and beautiful and as full of life. To show the complex beauty of a deciduous forest both in winter and in spring we chose the vast and lovely New Forest in Hampshire. The term 'new' is something of a misnomer since the forest was created by William the Conqueror in 1079, so in reality this great rustling belt of trees is nearly a thousand years old. It was a happy choice from my point of view, since for some years I lived not far from the forest and it became a wonderful hunting-ground for me with its rich and rare flora and fauna.

The forest today covers nearly 38,000 hectares, or about 80,000 acres. The larger part is open forest but there are also ancient and ornamental woodlands as well as grassland, heath and scrub. Originally of course the forest was created as a hunting preserve for royalty, though the local villagers had the right (and still have today) to graze their pigs, cattle, horses and other domestic animals over the forest except for certain enclosures that are created to protect young trees from the destructive attentions of deer and domestic beasts. However, now the forest is no longer a royal hunting preserve but, owing to its great importance and the many rare species that inhabit it, it has the status of a national nature reserve. Our guide to this lovely piece of forest, one of the finest in Europe, was Simon Davy, a tall, handsome young man, who was a great enthusiast and knew the forest intimately.

Naturally, Jonathan wanted to be as close to the forest as possible, for nothing is more aggravating or timewasting than having to drive for an hour every morning to get to your location. In this we were lucky, for Jonathan had discovered the Bramble Hill Hotel, lying in the middle of the forest itself. Whether the kindly but unfortunate owner, Captain Prowse, thought himself lucky to have us as guests is a moot point. We must have been a great trial to him, and by the time we left he must have considered that all film crews, if not actually certifiable, at least were considerably more eccentric than a set of village idiots. Certainly, our arrival and the incident of the bedspread got us off to a good start.

For some reason, Jonathan was for once considering the talent, and before we arrived he thought he would go up to our room and make sure that it was suitable for the stars. Why he would think that, in such an impeccably run hotel, anything would be amiss, I have no idea. Anyway, he found the double bed covered with a bedspread which, though maybe a trifle gaudy, was in no way offensive, but Jonathan became suddenly convinced that upon sighting this homely article Lee and I would be overcome with artistic rage of the sort that would have been expected from the late Lord Clarke if he had found some graffiti on the wall of Chartres Cathedral. Plucking the offending cover from the bed, he bundled it into the wardrobe. Thus having convinced himself that our delicate aesthetic sensibilities would not be offended, he set off to meet us. While he was so occupied, Captain Prowse – whose all-seeing eye was everywhere – had checked our rooms himself and finding the bed was, so to speak, naked searched for and found the bedspread and returned it to its rightful place. When we got to the hotel Jonathan, still suffering from that concern for the talent that so rarely afflicts a director, carried some of our bags up to our room ahead of us and saw the offending coverlet glaring at him from the bed. With a moan of anguish, he tore it from the bed and enshrined it once more in the cupboard. He was only just in time, for at that moment Lee and I entered the room accompanied by Captain Prowse. The Captain's eagle eye swept the room and came to rest on the bed. An expression of disbelief crossed his face.

'Why, where's the bedspread?' he enquired. A rhetorical question, but one that Jonathan thought he should respond to.

'Bedspread?' he croaked.

'Yes,' said Captain Prowse tersely, 'there was a bedspread on the bed. I put it there myself. Someone, for some strange reason, had put it in the cupboard. I wonder where it has gone?'

'It's in the cupboard,' said Jonathan in a low voice.

'In the cupboard?' said the Captain. 'Again?'

'Yes,' said Jonathan.

'How do you know?' asked the Captain.

'I put it there,' said Jonathan, with the demeanour of one confessing to infanticide.

'You put it there?' asked the Captain.

'Yes,' said Jonathan miserably.

'Did you put it there the first time?' As a military man, he quite rightly liked to be apprised of all the facts.

'Yes,' said Jonathan.

'Why?' asked the Captain, with ominous calm. There was a long silence while we all looked at Jonathan, who managed to achieve a rich blush that any self-respecting heliotrope would have been proud to wear.

'Because I thought they wouldn't like it,' he said, thus passing the buck to my poor innocent wife and myself. However, the Captain, with experience of slovenly recruits absent without leave and with a fund of plausible stories up their sleeves, was not to be distracted with this prevarication.

'Surely,' he said icily, 'if Mr and Mrs Durrell were in any way dissatisfied with their counterpane or bedspread, it was for them to inform me and not your place to secrete it in the cupboard. However, I have no doubt that Mr and Mrs Durrell will make up their own minds as to the suitability of the bedspread and communicate their displeasure or otherwise to me, without the intervention of a third party.' With that, he inclined his head and left the room – not a moment too soon, since Lee and I fell upon the bedspreadless bed in paroxysms of mirth.

It was mid-autumn, and the forest, when we started work in the morning, was looking magnificent. Parts of it were still shimmering green-gold, but in other parts the leaves were dying and the great trees stood barley-sugar gold, tawny as sherry or fox red in the fragile early-morning sunlight, with thin wisps of mist moving like kite tails through their branches. The air was cold enough to see one's breath, and everything had a delicate sheen of dew on it. Tiny streams glinted and whispered surreptitiously between the high banks of earth, as black and fragrant as Christmas cake, as they wended their way twisting and turning through the cathedral naves of the giant oaks and beech-trees.

Now, with the dampness in the air and the rich, moist layer of dead leaves, this was the time of the fungi and they were everywhere in profusion. Their fantastic shapes were like a Martian world. They seemed endless in form and colour. Mushrooms as pink as sugar icing, mushrooms grey and silky as a seal, mushrooms curved upwards, showing their gills like the leaves of a book, others like umbrellas disembowelled by the wind, some like dainty summer parasols, some like Chinese hats, others crowded together like tables in front of a French café or bubbling like waterfalls from the bark of trees. There were some like complex pieces of coral or slivers of orange peel; the sulphur-tufts, yellow as canaries, the grisettes a rich foxy red; parasol mushrooms, pale caramel-coloured with scales on top like tiles on a roof, or the pholiota, pale brown with scales like fur.

Then there were the fascinating names they were called. The scientists who collect and classify fungi have obviously a strong poetic streak in them and have given them names like the Shaggy Ink Cap, or Lawyer's Wig, the Weeping Widow, the Penny Bun, Slippery Jack and Dayrads Saddle. Then lurking among the trees you find the Varnished Death Cap and the ivory-white Destroying Angel, its uneven top looking like a tombstone angel's wings. Then there were the huge, flat, plate-like beefsteak fungi, clamped so tightly to the tree trunks it was possible to sit on them as one would sit on a shooting-stick without breaking them off. There were the puffballs, round and soft, which, at the touch of a finger, would give off a puff of minute spores, a silent, mist-like explosion that would send future generations drifting across the forest floor like trails of smoke. In one remote glade in the forest, we came upon the carcass of an immense oak which, at its death, must have been hundreds of years old, for the diameter of its trunk was all of ten feet. This enormous corpse must have been gently decaying there for some time and it had been transformed from a simple tree trunk into a complex fungi-garden. Crowds, consortiums, batches, battalions, clusters and caravanserai made up the biggest collection of different fungi I had ever seen since I was in a forest in Jujuy in northern Argentina many years ago. In fact, in my ignorance, I thought that only the tropics could produce this profusion of species in such a small area.

But of all the fungi we saw I think the one that took my fancy the most was the fly agaric, the size of scarlet saucers, lighting up the gloom under the trees. This gaudy thing, flamboyant as a trumpet blast, is poisonous, as has been known since medieval times, when the good housewife in her dairy or kitchen would put down saucers of milk with fragments of the mushroom broken into it to poison flies. Its poisonous properties lie in producing a cataleptic state accompanied by convulsions and a form of intoxication. Curiously enough, reindeer apparently adore it, treating it rather as we would treat a bottle of whisky or gin found under a tree and, I regret to say, when fly agaric is in season lose no opportunity to get plastered. The Lapps, who watched this performance and had seen the reindeer eat the mushrooms and possibly envied their regrettable state of intoxication, by experiment found out two interesting things. You could eat the fly agaric if you swallowed it without chewing, and so this is what they did to get the desired effect. They also found out – one shudders to think how – that if you imbibe the urine of someone who has been on an agaric bender you can obtain the same effect through this now 'distilled' potion. The Lapps of course, when they are blundering about with colossal hangovers, blame the reindeer for the whole thing.

Although to us the forest was ravishing, to Jonathan it represented an implacable enemy who, with its ever-changing moods, was trying to frustrate him. If he wanted sun, there was too much cloud; if he wanted cloud, there was too much sun; if he wanted rain, it remained blissfully clear, and vice versa. To us, the poor forest was doing its best in the only way it knew how. To Jonathan it was behaving with the malicious contrariness of a flibbertigibbet woman in a gown of multicoloured leaves. It was, in fact, the question of leaves that made him almost apoplectic. Since we were purporting to shoot the forest in autumn, Jonathan was not satisfied that there were vast quantities of dead leaves on the ground and an equal quantity up in the trees awaiting their turn to fall. He wanted pictures of leaves actually falling. Here, again, the forest exhibited its maddening, capricious, feminine nature. She provided him with huge piles of fallen leaves and towering tree-top pinnacles of shimmering green-gold, russet and chestnut-coloured leaves, none of which she would allow to fall – not to camera, at any rate. Any time the camera was safely packed away they fell in never-ending battalions, but the moment the camera was set up the leaves remained steadfastly glued to their branches. We began to fear for Jonathan's sanity and then the day came when we feared we might have to certify him.

'I've got it,' he barked at Paula. 'I've got it.'

'What, honey?' said Paula, noting the mad glint in his eye.

'Plastic bags,' said Jonathan triumphantly. 'You must go into town and get me several enormous, big, huge plastic bags.'

'Sure, honey,' said Paula soothingly. 'Anything you like – but what for?'

'Leaves,' said Jonathan. We all looked at him. He wasn't actually frothing at the mouth, so we decided to humour him.

'What have plastic bags to do with leaves?' I asked, not for a moment expecting a rational answer. I did not get one.

'We collect the leaves in the bags and take them back to the hotel,' said Jonathan.

'What do we do with them when we've got them back at the hotel?' asked Lee, fascinated.

'Dry them.'

'Dry them?'

'Yes, dry them, and then we get a ladder and climb a tree and throw them down,' said Jonathan. 'In that way I can get pictures of falling autumn leaves.'

This Napoleonic plan naturally brought us once again into direct conflict with the long-suffering Captain Prowse. Paula was dispatched to the nearest village and returned with four huge, funereal black plastic bags. Urged on by Jonathan, now delirious with excitement, we stuffed these with sodden leaves and returned to the Bramble Hill Hotel carrying with us enough potential humus to succour the world's major botanical gardens. We grouped this largesse in the foyer of the hotel and Jonathan went in search of Captain Prowse. When they returned, he showed the Captain the four huge plastic bags, their shape distorted by their contents, looking like malevolent slug-shaped things from outer space.

'I want your help with these,' said Jonathan simply. The Captain examined the bags with care.

'With these?' he enquired at length. 'You want my help?'

'Yes,' said Jonathan.

'What are they?' enquired the Captain.

'Leaves,' said Jonathan.

'Leaves? What sort of leaves?' asked the Captain. Nothing like this had ever happened to him before.

'Autumn leaves,' said Jonathan triumphantly. 'We got them in the forest.'

Captain Prowse looked shattered. Nothing in his previous career had prepared him for a guest suddenly producing four plastic bags full of autumn leaves and demanding his help.

'I see,' he said, moistening his lips. 'And what do you intend to do with them?'

'Dry them,' said Jonathan, puzzled that the Captain could not have worked out this logical progression for himself.

'Dry them?' asked the Captain. 'Dry them?'

'Yes, they're wet,' Jonathan explained.

'But why should you want to dry them?' asked the Captain, fascinated in spite of himself.

'Because they won't fall if they're wet,' said Jonathan, impatient at the Captain's obtuseness.

'But they have already fallen,' the Captain pointed out.

'I know,' said Jonathan, exasperated. 'That is how they got wet and that is why we have to dry them.'

Mercifully, at that moment Paula, who had been off making one of her frequent telephone calls, reappeared. She took in the situation in one sweeping glance.

'Captain Prowse, I think perhaps I ought to explain and I am sure you can help us if anyone can,' she said, smiling and turning on the 5,000 candlepower of her charm.

'I would be glad of some clarification,' said the Captain.

Simply and concisely, Paula explained the whole drama of the falling leaves. Initially, Captain Prowse had kindly put at our disposal a room (in addition to our bedrooms) in which we could assemble for script conferences, and which would allow us to spread out and maintain the equipment. It was a strange room on the first floor, rather like a Victorian conservatory. Now Paula asked simply if we might also dry half the New Forest in it. It says much for the Captain's good-humour and his firm grasp of being a good hotelier that he did not immediately ask us to leave. Instead, he gave us a huge pile of back copies of *The Times* on which to spread our largesse of leaves and a large and formidable electric fire, *circa* 1935, with which to dry them. Soon the fire was throwing out heat like a blast furnace and the leaves were spread out on *The Times* occupying half the room, while Jonathan crooned over them, stirring them lovingly with his fingers. We all foregathered to drink whisky and watch him.

'It looks like a village-hall setting for a panto,' said Chris. '*Babes in the Wood*, perhaps.'

'No,' I said judiciously, 'Harris is too old for a babe. It's more like *The Tempest*. There's old Caliban, groping about in his spiritual home.'

'You may laugh,' interrupted Jonathan coldly, continuing to stir his leaves with loving fingers, 'but you wait until we have real autumn leaves raining down from the trees.'

Two days later, when the leaves were dry, we carried them reverently out into the forest. With them we solemnly transported a ladder and, under Jonathan's direction, propped it against the trunk of a huge oak tree. Brian, who was not doing any sound-recording in this sequence, was detailed to go up the ladder carrying a plastic bag full of leaves, crawl out along the branches and start scattering leaves as though he were Mother Nature. This he dutifully did.

'Throw them more naturally,' Jonathan kept calling.

'How can you throw them naturally out of a plastic bag?' asked Brian aggrievedly from his precarious perch.

'Throw them delicately,' said Jonathan, 'not in great wodges like that.'

'I must say, you chaps go to an awful lot of trouble,' said Simon.

'No expense spared,' I said. 'Erich von Stroheim, when he was filming, once had 35,000 almond blossoms fixed to the trees because he was shooting in winter and the trees were bare.'

'Good God, wasn't that expensive?' asked Simon.

'Yes,' I said, 'very. Harris is related to him, of course, hence this leaves business.'

'Really?' said Simon, interested.

'Yes, his real name is Harris von Stroheim, but he changed it.'

'So that is why he is so keen on the leaves?' asked Simon.

'Yes, well, with our budget you can't run to almond blossom,' I said.

As I have said, Jonathan felt that the New Forest was not

really co-operating. It deliberately grew its fungi in shady corners with not enough light for photography. It refused to shed its leaves, it got rained on, it got covered in fog, it was recalcitrant to a degree. Then came the final straw, the business of the galls.

Each tree in a forest has, of course, an ecosystem of its own. The tree itself, while controlling heat and moisture and so climate, provides an important world for a host of creatures who live on, in or around it or merely visit it for reasons of business, like nesting. It has been estimated that a single oak tree can support well over three hundred species (and goodness knows how many individuals of each species), ranging from birds to moths, from caterpillars to spiders. Among the creatures to which the oak tree is a world in itself are the many species of gall. Galls are some of the most bizarre and decorative things you can find in a forest, and Jonathan had been much struck by what I had written about them in *The Amateur Naturalist*. I had said:

> Each gall forms a home for a developing larva. In some the adult insect hatches out in the summer, in others the galls turns brown and the larva hibernates through the winter inside it. But the story of the galls does not end there, because within each gall you will almost certainly find other creatures which are either acting as parasites on the original owner-builder of the gall or who have just taken up residence as unpaying guests. The common oak-apple, a very easy-to-find gall, has been known to give a home to 75 different species of insect as well as the rightful owner, the gall-wasp grub.

It was that phrase, 'the common oak-apple, a very easy-to-find gall', that did it. Jonathan was determined to obtain some of the oak-apples and film Lee and myself collecting them and then take them to London Scientific Films (who were doing all our close-up photography for us) and get the galls to hatch out while microphotography captured every one of the seventy-five different species in gigantic close-up. Now, normally, in any good forest, you – as it were – can't see the wood for the galls, but on this occasion things were different. We set off early one morning on a gall hunt, Jonathan bearing two of his gigantic black plastic bags which only recently had contained leaves.

'Are you sure two will be enough?' I asked.

'You said they were common and easy to find,' he said. 'I want lots of them.'

'Well, that bag will hold two thousand at a rough guess, so the two of them will cope with maybe four or four and a half thousand.'

'I don't care,' said Jonathan stubbornly. 'I'm not taking any chances; I want lots.'

So we set off into the forest, like a lot of pigs in search of truffles.

First, we started on the baby oaks at the edge of the forest proper. These small trees were always favourites with gall-makers and from our point of view, being small trees, were easier to examine. We must have examined several hundred. Not only were there no oak-apples, there were no galls at all. Jonathan was getting restive, as he always did when nature refused to obey him.

'Hey, guys,' bellowed Paula from about a quarter of a mile away, making one's eardrums vibrate, even at that distance. 'Say again what they're like.'

'Like small brown rotting apples,' I yelled back.

We searched on. We left the smaller trees, each glossy and gall-less, and moved into the forest towards the taller trees. We had started at eight and by eleven I began to believe that the forest was bewitched; Jonathan, I decided, had cast a spell on it. In all my experience, I had never come across anything like it. I had never *been* in an oak forest without finding oak-apples. It was like going to the Sahara and finding no sand. Then at about eleven-thirty Lee gave wild cries of delight.

'I've found one. I've found one,' she yelled.

We all converged on her at a run.

'Where, where, where is it?' barked Jonathan hoarsely. Lee pointed. Attached to the branch of an oak-tree she was holding was an oak-apple. It was undeniably an oak-apple but so minuscule, so shrivelled, so pathetic, that it looked like a very old dropping of a Lilliputian elephant.

'Is *that* an oak-apple?' asked Jonathan suspiciously.

'Yes,' I said, 'but I must admit I've seen healthier specimens.'

'Well, it's the only one we have,' he said, as he removed it carefully from the tree. 'We'd better take it.'

As it turned out, it was the only one we did see. It was transported back to London as reverently as if it had been the Crown Jewels (or, rather, Crown Jewel), and people sat around it for weeks, cameras at the ready, like scientists waiting for something to emerge from a flying saucer. Nothing happened. When it became patently obvious that nothing was going to hatch, Jonathan cut it in half with a penknife. Inside was one very small, extremely dead larva of a gall wasp. Filming nature is not easy, especially when you only have a limited time to do it in.

The next thing that we had a slight contretemps with were badgers, those magnificent creatures of ancient lineage who have waddled through the English woodland since the days when humans were dressed in woad. It is a swaggering, beguiling, beautifully designed creature of great intelligence and charm which does an enormous amount of good as one of the countryside's major predators, striking terror into the heart of everything from a woodlouse to a baby rabbit, a pheasant chick to a frog, taking in worms, snails, beetles, snakes and hedgehogs on the way. The word 'omnivorous' means 'eating everything', and the badger lives up to this title admirably. Everything is grist to his mill. In spite of this indiscriminately carnivorous approach to life, much of the badger's food consists of roots, mushrooms, berries and seeds. Altogether, it is a handsome and useful addition to the countryside and if it does occasionally wreak havoc in a cornfield or a hop field, or set a henhouse on end, one must overlook these lapses from good manners for the amount of good these creatures contribute.

Badger homes, or setts, are enormous complex structures of endless tunnels and chambers. As the sett (like the English country home) is handed down from generation to generation and as each generation enlarges and improves it, the ramifications of an old sett are considerable. It consists of bedrooms, recesses and, where young are being reared, they even excavate special lavatory areas. Badgers mate for life and, being eminently civilized beasts, remain on good terms with all the neighbouring badger couples.

Just recently, the badger, who has been shambling through the green twilight of the English woodland for a millennium, has been beset by two separate groups of so-called civilized man. They were accused of carrying bovine TB (which they probably do) by that group of veterinary surgeons employed by the Ministry of Agriculture. Their answer to the problem was 'Kill the Badgers' and so there was a great flurry of badger-gassing under the most unpleasant conditions. It has always seemed to me that veterinary surgeons employed by the Ministry have only one answer to any problem, which is 'destroy it' rather than 'solve it'. Fortunately, this simplistic approach was stopped by public outcry, and the vandalizing of the ancestral homes of the badgers and the destruction of this creature was halted. You would have thought that an official campaign of gassing (macabrely Teutonic in its conception) would have been quite sufficient for the poor badger, but no. Once officialdom had been worsted, the animal was threatened on another front. Badger-baiting with terriers became the lead sport among those members of the human race whose frontal lobes are still Neanderthal. Badgers, who probably do more good per annum to the environment than these gothic human horrors, were dug out and then beset with terriers. This persecution of the badger represents the two ends of our rainbow of society. The bureaucratic manipulation at one end and the wonderful, much lauded working man at the other, who, because he likes (like a Roman crowd) a little bit of blood-letting, drags society with him, sweating and grunting in his pursuit of pain and death.

We had already obtained marvellous footage of badgers underground, filmed for us by Eric Ashby who has allowed badgers to build a sett under his cottage. With the aid of two-way glass he can observe and film his badgers underground. To complete this badger sequence, what Jonathan now wanted was some shots of Lee and myself outside a badger sett and the animal emerging in front of us.

'You will be sitting outside a sett on the other side of the valley,' Jonathan explained, 'and then, just as it becomes dusk, the badgers will come out.'

'Have you told the badgers this?' I asked.

'They will come out,' Jonathan said confidently. 'They will come out for the sandwich.'

'Sandwich? What sandwich?' asked Lee.

'A peanut-butter sandwich,' said Jonathan.

'What are you talking about?' asked Lee.

'Badgers,' said Jonathan, with an air of authority, 'badgers find peanut-butter sandwiches irresistible. They will travel miles to obtain one. Drag a peanut-butter sandwich through the forest and you will have every badger for miles around following you.'

'Where did you obtain this esoteric piece of information?' I enquired.

'I read it in a book on badgers,' said Jonathan. 'They said it never fails.'

'It sounds distinctly peculiar to me,' I said. 'I have never heard of attracting badgers with peanut butter.'

'Chipmunks like peanut butter,' said Lee. 'I used to feed it to them in the garden in Memphis, so I don't see why a badger shouldn't like it.'

'They find it irresistible,' repeated Jonathan. 'They'll do anything for a peanut-butter sandwich.'

So, armed with a vast quantity of peanut-butter sandwiches, we made our way into the forest where there was an extensive badger sett. However, it was a fairly well-populated area and all the signs were that a number of human beings had been trampling round in the vicinity.

'I hate to sound pessimistic,' I said to Jonathan, 'but what happens if the badgers won't come out on cue?'

'I've thought of that,' he said, glancing at his watch. 'Any minute now my badger reinforcements will arrive.'

'What reinforcements?' asked Lee.

'A chap called David Chaffe,' said Jonathan. 'He has two tame badgers. He will be arriving with them any minute now, so if the wild ones don't show up we at least have some tame ones.'

So the cameras were set up, and Lee and I took up our badger-watching positions. Needless to say, no badgers appeared. It was not altogether surprising as, however careful a film crew are, they cannot be completely silent and badgers have very sensitive hearing. By now, however, David Chaffe, burly and bearded, had arrived and in his van were two handsome young badgers. These, wriggling and snorting with excitement, were removed from their cages and released outside the badger sett.

'Now,' said Jonathan, directing, 'what I want you to do is to say to camera: "There is one sure way of attracting a badger out of its sett. It shares with us humans a passion for peanut-butter sandwiches, and using one as bait you can get the most recalcitrant badger to show itself." Then you throw the sandwich just outside the entrance to the sett and the badgers will fall on it with delight.'

Dutifully, I did my piece to camera and then threw the sandwich at the entrance to the sett. As on cue, the two badgers approached it, they both sniffed the sandwich and then backed hurriedly away from it, sneezing violently and displaying every symptom of acute displeasure. It was more than obvious that a peanut-butter sandwich was not considered a delicacy by this pair of badgers. And of course this incident went towards strengthening Jonathan's opinion that Mother Nature was just not co-operative in any way at all.

SHOOT SEVEN

HAVING COLLECTED ANIMALS all over the world, I am always amazed at the rich variety of wildlife you can find in an ordinary hedge in the English countryside.

For over a thousand years, the cattle and crops in English fields have been guarded by the thick English hedgerows. When the Saxons first invaded Britain, they started felling the forest, or Wild Wood as it was called, in order to create clearings for pasture or crops. In order to demarcate these and prevent cattle from straying, the hedgerow came into being. The Saxons soon discovered the ideal plant with which to construct hedgerows – the hawthorn. Easy to grow by layering or simply planting out branches, the hawthorn would rapidly blossom into a thick, impenetrable, spiky hedge, impervious to cattle, a perfect windbreak and providing a mini-forest for wildlife to replace the great forests that had been felled. It is estimated that something like half a million miles of hedgerow lay like a chessboard across the countryside. As well as guarding his crops and cattle and providing wildlife with a refuge, the hedgerows were a veritable apothecary's shop to medieval man, providing him with many plants to cure everything from headaches to hernias and other plants to stave off the meddling of witches, so prevalent in those days. The lovely stitchwort, white as a snowflake, was supposed to cure sharp pains in your side; the cool, fleshy dock leaves were soothing for stings, especially nettle stings; if you were wounded or cut, the self-heal would be the plant to search for; or, if you suffered from ulcers in the mouth or in more intimate and painful places, the handsome silver weed would cure you. Medieval man treated the hedgerows with respect because they were the haunt of elves, fairies and other sprites. As well as bad magic lurking in the hedges (if you plucked lady's smock an adder would bite you, or pluck innocent blue speedwell and create a thunderstorm or, worse, bring a bird down to peck out your eyes), there was good magic as well. You could rub your cow's udder with buttercups and thus increase your milk yield or you could hang nettles in your dairy to prevent witches from curdling your milk.

Once the hedgerows were considered vitally important and they were carefully maintained so that man and wild creatures benefited from them. Now, however, farmers consider the hedgerow to be a nuisance and so these ancient and useful sanctuaries are being bulldozed out of existence to make way for ever larger and larger tracts of land, exposed to the eroding effects of wind and rain.

We wanted to try to show on film one of these ancient hedgerows that still exist and capture some of its beauty and importance. In time, the hedgerow will be a thing of the past and a lovely and important part of Britain's heritage will have vanished. So we wanted to show a bit of the English countryside as it had been for thousands of years and to explore it. We decided to use three archaic or semi-archaic forms of transport. Jonathan had found a magnificent area of hedgerow in Sussex which ran along the side of a sunken green lane, that is to say a natural lane, uncorrupted by gravel or macadam, the sort of lane a-glitter with flowers, sheltered by tall hawthorns covered with white blossom like snow. It was the sort of green lane that Shakespeare knew, the sort of lane that the pilgrims to Canterbury used. The first stage of our journey to this entrancing bit of English countryside was accomplished, to my delight, by steam train.

Railway enthusiasts have, in different parts of the British Isles, rescued ancient steam engines, lovingly restored them, and are allowed to run them on special sections of rail. You discover that the driver, the guard, the conductor and other personnel are in real life schoolmasters, professors, shopkeepers, chemists or merely retired train buffs who give their services free so that this generation can experience the feeling of real train travel, can smell that magic, acrid perfume of coal, soot and steam, can thrill to the owl-like hoot of the engine itself, can be sent into a trance by the rattle, clank and hiss of the train and then settle to that rhythmic clackety-clack like the heartbeat of the train itself. Eagerly, we wended our way down to the charmingly named Bluebell Railway.

Lee, at the age of thirty-three, had never travelled on a steam train – a revelation which shocked me to the core of my soul, railway enthusiast that I am. So we arrived at the station and there was the train, gleaming and glittering, wearing a rakish scarf of steam over one shoulder, and behind it the elegant carriages, quite rightly designated as First, Second and Third Class. The heavy doors slammed with a satisfying clunk and had great leather straps with which to lower the windows, the more easily to get sparks in your eyes or soot on your nose – all experiences that enhance any railway journey that *is* a railway journey. Jonathan, regardless of the budget, had thrown caution to the wind and reserved a whole First Class carriage for us, with wide, beautifully upholstered seats with fat buttons like mushrooms embedded in them. There were gaily coloured travel posters depicting seaside resorts *circa* 1920, with a sea of such blueness you wondered why anybody ever went to the Mediterranean. There were huge luggage-racks, wide and sturdy enough to take any number of Gladstone bags, hatboxes, picnic hampers and other necessities. Lee and I took our seats by the window, while Chris and Brian wedged themselves with the recorder and the camera into the further corner, for the start of the programme was to be me doing a voice to camera as the train chugged through the English countryside. For a variety of technical reasons, we had to shoot this scene several times, which was excellent as far as we were concerned, as it meant that we had to travel to and fro up the Bluebell Railway several times also, thus getting twice as long a journey as we would have had normally on this enchanting train.

Finally, however, the opening sequence was in the can and reluctantly we had to leave the train. It halted miles away from anywhere at a small wooden platform bearing a large white sign saying 'Freshfield Halt. Please signal if you wish the train to stop.' We got out on to this rickety platform and then extricated from the guard's van the next form of transport we were going to use on our journey – a large sparkling tandem bicycle.

This had been Jonathan's brilliant idea and, although I protested that the last time I had ridden a tandem was in 1939, Jonathan insisted as usual that it was simplicity itself. The machine had arrived at our hotel the previous evening, and Lee and I had decided to take it into the forecourt of the hotel and practise. We got off to a somewhat wobbly start, as Lee insisted that she knew how to pedal better than I did. Also the bicycle was extremely light and so you had to be very cautious with the turning or the machine folded up its front wheel like the wing of a bird and you found yourself in the ditch. However, we soon mastered it and were whizzing around the forecourt at a rate of knots when, unfortunately, three old ladies accompanied by someone who looked like a brigadier-general of the old school suddenly emerged from the hotel and tottered straight across in front of us. I crammed on the brakes, the tandem skidded on the gravel, the front wheel folded up, and Lee and I fell in a tangled heap, inextricably entwined with our machine. The three old ladies screamed, the Brigadier-General said something that sounded suspiciously like 'By Gad!' and Lee and I, rolling about on the gravel, finally managed to disentangle ourselves and stand up. The Brigadier-General screwed a monocle into one eye and surveyed us from top to toe. We were both wearing our naturalist outfits, which, after a day's filming in the rain, looked extremely scruffy.

'Trippers!' said the Brigadier-General, after surveying us in silence for a moment, packing into that one word all the well-bred scorn that the middle-class Englishman still feels for the proletariat. Then, putting his arms protectively round the old ladies, he ushered them away from contamination. It was not an auspicious start.

However, at Freshfield Halt, when the Black Knight, in a cloud of steam, had chuffed away uttering farewell whistles of a piercing clarity, the scents and sounds of the May countryside were wafted to us in the spring sunshine. Everywhere there were larks embroidering the blue sky with their song. Cuckoos called loudly and persistently in the fields and the scent of a hundred spring flowers filled the air. We manoeuvred Daisy, as we had christened our tandem, down the wooden ramp on to the cinder track and then down a narrow slippery path which led finally to a narrow lane with high banks covered with a glittering army of kingcups yellow as saffron, and the bank topped with hedges of hawthorn, their blossom like cumulus clouds. So, mounting Daisy and with the sun hot on our backs and the birdsong ringing in our ears, we set off in search of ancient England.

The piece of countryside which Jonathan had chosen could not have been more perfect or, at that time of year, more beautiful. The tall banks and hedges were glowing with a multicoloured embroidery of flowers, the canary yellow of buttercups, red of campion, white stars of stitchwort, the mist of bluebells and the lavender of violets, and the curious flat flower clusters of the cow parsley looking like pale smoke. The meadows between the hedges were huge and lush, sprinkled with flowers and stands of impressive oak and beech trees casting pools of shade with their newly emerging leaves. Where there were cottages or larger houses, these were discreetly hidden in belts and groves of trees, so that they were not obtrusive and one got the impression that the countryside was virtually uninhabited.

At last we came to the sunken green lane with its impressive hedgerow shielding it, at one side a thick, almost impenetrable wall of hawthorn interwoven with the odd oak, its roots covered with a web of ivy. Here we met up with Dave Streeter, who was to be our hedgerow guide, and an excellent one he proved to be. Slim and dark, he had the bright sprightliness of a bird, with his dark eyes and inquisitive, beak-like nose. He was as proud of the hedgerows as though he had planted them himself and knew every bird, insect and plant that inhabited them. With his aid, we unravelled the secrets of this ancient living wall.

Most hedgerows count their birthdays in centuries but naturalists have evolved a fairly simple method of working out the approximate age of a hedgerow. You measure out and mark thirty paces along it and then retrace your steps and count the number of woody plants growing along its length. Each one of these is the equivalent of a century. This may sound improbable, but it is based on some sound detective work by naturalists. When the hedge is first laid down, the farmer uses one or maybe two kinds of plant. Over the years, other plant species spring up, brought in the form of seeds in the droppings of birds and by squirrels and mice who bury nuts and seeds and then forget where they hid them. By working from hedges of a known age, it became apparent that the rate that woody plants become established is about one species per hundred years.

So Lee and I paced out a length of our hedgerow and then collected samples of all the different woody plants. We found over ten different kinds, which meant that, when this hedgerow had been laid down, the Tower of London and Westminster Abbey had still to be built. It is amazing, the reverence which these two buildings are accorded and yet the hedgerows of England which for over a thousand years have been doing so much good for man and nature are being bulldozed out of existence with only the faintest outcry against this unbiological brutality. If anyone suggested clearing away Westminster Abbey to make way for an office block or wiping out the Tower of London to make way for a new Hilton, the outcry would be fantastic, yet in the period of a thousand years that these two buildings have been in existence they have been probably of considerably less use to mankind than the humble hedgerows.

Quite apart from the many plants that nestle at its feet or climb up into its spiky canopy, the hedgerow provides a home for a great variety of reptiles, birds and mammals, some of which we managed to film. One of the most attractive from my point of view was the harvest mouse, the most diminutive mammal in the British Isles and one, moreover, who has the distinction of having been discovered by no less an amateur naturalist than the great Gilbert White himself. In his enchanting *Natural History of Selborne* he notes the harvest mouse's arrival into scientific ken thus:

I have procured some of the mice mentioned in my former letters, a young one and a female with young, both of which I have preserved in brandy. From the colour, shape, size, and manner of nesting, I make no doubt but that the species is nondescript. They are much smaller, and more slender, than the *Mus domesticus medius* of Ray; and have more of the squirrel or dormouse colour; their belly is white, a straight line along their sides divides the shades of their back and belly. They never enter into houses; are carried into ricks and barns with the sheaves; abound in harvest; and build their nests amidst the straws of the corn above the ground, and sometimes in thistles. They breed as many as eight at a litter, in a little round nest composed of the blades of grass or wheat.

One of these nests I procured this autumn, most artificially platted, and composed of the blades of wheat, perfectly round, and about the size of a cricket-ball; with the aperture so ingeniously closed, that there was no discovering to what part it belonged. It was so compact and well filled, that it would roll across the table without being discomposed, though it contained eight little mice that were naked and blind. As this nest was perfectly full, how could the dam come at her litter respectively so as to administer a teat to each? Perhaps she opens different places for that purpose, adjusting them again when the business is over; but she could not possibly be contained herself in the ball with her young, which moreover would be daily increasing in bulk. This wonderful procreant cradle, an elegant instance of the efforts of instinct, was found in a wheat-field suspended in the head of a thistle.

The harvest mouse has adapted itself to an arboreal life in the same way as many of the New World primates have done. It has agile feet for gripping the grass stalks it lives amongst and has developed a prehensile tail of astonishing power by which it can actually hang from a grass stalk while building its nest. These round nests, roughly the size of a tennis ball, are woven for the most part out of living grass blades, occasionally being reinforced with chewed-off leaves. These nurseries – for this is where the female rears her young – have two entrances and are lined with finely chewed grass blades to form a soft bed for the young. The babies when they are born weigh about a gram or so; as Gilbert White observed, two would be the weight of a copper halfpenny. A new nest is built for each litter, and in a good year the harvest mouse is capable of giving birth to six litters of five or six young per litter. This sounds like gross overpopulation as practised by the human race. However, it must be pointed out that, when there is a glut of harvest mice, the creatures that prey on them, such as foxes, weasels, stoats, owls and so on, have a field day and as a rule increase their own families enormously. When the harvest mice have a bad year and don't overpopulate, the predators have a hard time and so their families are regulated to the mouse supply. It is unfortunate that mankind now only has one predator – himself. But his overpopulation is so great that the predation of his own species does not keep the population in balance in the same efficient way that nature does it.

Another occupant of the hedgerows is the hedgehog. These have always been favourite animals of mine ever since, during my childhood in Greece, I hand-reared a litter of four, brought to me by a peasant who had dug them up in their leafy bed at the edge of his field. Newly born hedgehogs are creamy-white in colour and their spines are quite soft, like india-rubber. As my babies grew, they gradually changed colour to brown and their spines became hard and sharp. They were, I found, remarkably intelligent little animals, and I even managed to train mine to stand on their hind legs and beg for scraps of food. I used to take them for long walks in the countryside and they would trot along at my heels in an obedient line. They were incredibly quick, and when I turned over logs or stones in search of specimens for my collection I had to be on the lookout, for they would rush in and scrunch up my coveted insects if I did not watch them. One day, they were foraging around some old vine stumps and I, finding the open vineyard hot, made my way to the cool shade of the olive groves about a hundred yards away and sat down. I could see my hedgehogs but they could not see me. It was some little time before they realized that I had disappeared and they were immediately filled with alarm and consternation. They ran round and round in circles, squeaking plaintively to each other; then one, nose to the ground like a dog, found my scent and set off at a brisk trot, leading the others. That it was undoubtedly my scent they were following became apparent, for I had meandered to and fro to reach my present position and they followed the route I had taken slavishly. There was much excited noise when they discovered me and they clambered squeaking and snorting into my lap.

I remember once, when we lived in Hampshire, we had a huge and ancient cooking-apple tree in the garden. One year it had a bumper crop of fruit, more than my mother could utilize in spite of making tons of chutney and jams, so a lot of fruit fell to the ground and we let it rot and manure the earth beneath the tree. One bright moonlight night, I was woken by screams and squeals and grunts and, thinking it was a pair of courting cats, I leant out of the window to give them a piece of my mind and saw to my astonishment that it was a pair of hedgehogs. Wondering what on earth they were doing, I donned my slippers and went down into the moonlit garden. I discovered that they had been feasting on the semi-rotten apples and the fermenting fruit had acted like cider, so now both hedgehogs were appallingly inebriated. They staggered round and round the tree bumping into things, hiccuping, hurling abuse at each other and generally behaving in the most reprehensible manner. For their own good, I had to lock them up in our garage overnight and the next morning it was a very dejected and sorry-looking pair of insectivores I released into the woods at the back of our house.

Another creature that we were lucky enough to film was the weasel, the smallest and most delightful of the British predators. Some twenty-eight centimetres in length, including their tails, they are beautiful, swift and slender little creatures – how swift we soon discovered when we started to film. In order to get the close-up shots of our weasel hunting, we built a very lifelike set to represent a section of the hedgerow. A film is made at twenty-four frames a second, that is to say that the camera takes twenty-four photographs each second. We found our weasel was moving so fast that he could actually cross the set *in between* the photographs – an absolutely extraordinary feat of agility.

I remember when I worked as a student keeper at Whipsnade I used to cycle on my day off across to Tring Museum to take lessons in taxidermy. On the way there was an ancient caravan and in it lived an old gypsy whom I frequently used to visit, since he kept innumerable pets and was always adding to them. My attention was drawn to this old man – whom everyone called Jethro – when I was cycling past his caravan and was suddenly riveted by the sight of no less than five weasels gambolling about the caravan wheels. I dismounted and watched them playing Catch as Catch Can, their bodies so sinewy they were like furry snakes. Old Jethro appeared out of the woods, a gun under his arm, carrying two dead rabbits. He gave a musical whistle and the weasels stopped their game and rippled across the grass to his feet, standing on their hind legs and uttering little yarring cries. He dropped the rabbits and the weasels, snarling and fighting, dragged the corpses under the caravan and started to feed. How I coveted those sinewy, delicious creatures but, alas, old Jethro would not sell them, even though I offered him my week's salary of £3 10*s* (£3.50).

'No, I won't part, boy,' he said, watching the weasels affectionately with his bright black eyes. 'Not with the trouble I had a-rearing of them. No, I wouldn't part, not for all the tea in China, but I tell you what, I'll take 'ee out a-hunting with them. Braw little hunters they be an' all.'

So one summer's night, when the moon was as full and white and round as a magnolia blossom, I cycled over to old Jethro's caravan. After a pint of beer (home brewed) and a delicious plate of stew, we set off, the weasels rippling ahead of us along the hedgerows, bathed in moonlight as bright as day. The weasels had evolved their own particular hunting method, old Jethro explained. One or sometimes two of them would enter the rabbit's burrow and the others would wait outside. Presently, panicked by the two weasels underground, the rabbit would bolt out of its burrow and straight into the group of weasels outside. They would converge on it like lightning and one of the three would dispatch it with the characteristic weasel bite at the back of the skull, the lower teeth driving upwards and the upper canines sliding downwards into the brain. Death was instantaneous. It was wonderful to watch the weasels dance snakelike in the moonlight, eyes occasionally gleaming, working as a team, lithe and silent. Whether they hunt like this in the wild is a moot point, but certainly these hand-reared ones had evolved a co-operative hunting method which was as efficient as it was deadly and within two hours old Jethro's poacher bag contained the bodies of seven fat rabbits. Some would be used to feed the weasels and his other meat-eating pets (he had owls and hawks, a badger and a stoat, among others) and the rest he would eat or take to the nearest village and sell.

Old Jethro used the hedgerows around his caravan as medieval man used them – to provide himself with food in the shape of rabbits or partridges, various herbs and roots to flavour the food, and other herbs which he made up into ointments and salves which he used to peddle in the local market towns; and I knew several people who would not go near a doctor but took all their ailments to him. I had a girlfriend who used to suffer acutely with an unsightly rash which would break out periodically on her forehead and in the palm of her left hand and irritate exceedingly. In spite of her protestations and disbelief, I took her to old Jethro and forced her to use the ointment he gave her. Three applications and the rash vanished, never to reappear.

In one of the final scenes of the programme, Jonathan wanted to show an ancient meadow of the sort that hedgerows have guarded for centuries. When he led us to the meadow of his choice we were delighted. It was vast, and guarded on three sides by tall hedgerows and on the fourth side by a dark piece of woodland, glittering with spring leaves. It sloped gently into the sun, with here and there a few oaks that, from their girth, must have been several centuries old, casting pools of bluish shadow. But the real breathtaking thing about the meadow was its colour. The long, lush grass was bejewelled with buttercups of such flamboyant richness it looked as though someone had, from some vast celestial vat, poured molten gold between the wood and the hedgerows. We had to walk to the centre of this field of the cloth of gold to have our picnic and it seemed sacrilege to wade calf-deep through the buttercups, leaving a crushed path behind us across that impeccably unsullied sheet of gold and green.

In the final scenes of the programme, in order to show the complex web of hedgerows spreading across the countryside, Jonathan had decided to send us up in a hot-air balloon. Although I had often wanted to try this splendid, archaic form of transport, I was a trifle nervous because of my vertigo. However, this was an opportunity too good to be missed, so I agreed to try to curb my absurd complaint and take to the air. The whole thing had to be planned like a military operation. We were to take two flights; the first day we would be accompanied by Chris and the camera so that he could get all the close-ups of us in the basket, while the others followed us by car and filmed us on the ground. On the second day, Chris and the camera were to follow our flight in a helicopter – and a helicopter to be piloted by Captain John Crewdson, no less, who had done all the complex and risky filming for the James Bond films. The pilot of our balloon was Jeff Westley, a skilled balloonist who could virtually land his craft on a sixpence. Ideally, Jonathan would have liked to have had the balloon ascending majestically out of the centre of the buttercup field, but this would have created too much havoc on the flowers and the grass, so we had to choose a much more plebeian and well-grazed pasture for our first ascent.

We arrived at the meadow early in the morning to find our balloon awaiting us. It was a gorgeous monster – far bigger than I had expected – gaily striped in red, yellow and blue. Resting on the grass beneath it was the basket, rather like a giant-sized old-fashioned laundry-basket, which contained the essential canisters of butane gas which made flight possible. We were introduced to Jeff, a stocky, fair-haired man with twinkling blue eyes and a massive air of confidence. He informed us that the weather forecasts were excellent and he looked forward to giving us a splendid flight. In order to get the necessary close-ups of us in the basket, Chris was to travel with us with one camera, but to cover wide-angle shots of the basket the camera had to be some distance away. We solved this problem by mounting a remote-controlled camera on a long aluminium beam, which could be operated from within the basket. Jonathan was anxious to make it appear as though Lee and I were 'driving' the balloon ourselves, so we took with us a large blanket and every time we were shooting – Jonathan explained to Jeff – he would have to crouch down in the bottom of the basket and we would cover him with the blanket. He agreed to be subjected to this indignity with great good-humour. So, with these last-minute instructions from Jonathan, we scrambled into the basket and prepared for our first ever balloon ascent. The anchor ropes were cast off and the basket shifted slightly, then Jeff pulled the cord and a giant tongue of blue flame above our heads roared into the interior of the balloon with a great blast. It was like unleashing a dragon. Aided by these deafening blasts, the basket rose smoothly as a lift. We glided up twenty feet, thirty feet and then slid up into the sky above the trees.

The sensation was miraculous. When the flame was not roaring the silence was complete, and a thousand feet below you as you wind-drifted smoothly through the sky you could hear people talking. You could hear the clatter of a train, a dog barking, or cattle lowing. I can only compare it to the sensation you get snorkelling on a tropical reef, where you can lie face downwards in the buoyant waters and let the tiny eddies of water drift you over the coral-gardens. Far below us, the patchwork quilt of fields, guarded by their hedgerows, stretched as far as you could see, with here and there a dark reef of woodland and here and there a toy village. Our shadow, like a great blue mushroom, glided over the fields and hedgerows below us, overexciting herds of cattle and making horses behave as if they were in a rodeo. Although you are at the mercy of the wind, there is a lot you can do to help steer the balloon, as Jeff showed us. At one point, the wind velocity dropped and he took the balloon down to tree-top height. We drifted along silently and gently as mist, and at one point the bottom of the basket actually rustled its way through the topmost branches of a gigantic oak tree. We saw a hare and any number of rabbits who found the presence of our fat, highly coloured vehicle alarming. We saw a pair of roe deer, standing prick-eared and tense in a woodland glade, and we were treated with vociferous rage when we slid over a rookery, so indignant were they at this untoward invasion of their airspace.

It was fascinating to drift some fifty feet up over the villages and isolated farms and cottages, for you could see everyone's back garden, beautifully tended and with a riot of flowers. The roar of our balloon would set all the dogs barking hysterically and people would run out of their houses to wave to us. As soon as they realized we could hear them quite clearly and reply, they would ask us where we were going, and were vastly amused when we said we did not know. We drifted over a village school, and all the children and their teacher tumbled out into the school yard to watch our progress. Inevitably, the children shouted up to us and asked where we were going. Inevitably, we replied we did not know. The children found this rather funny, one little boy laughing so heartily that he fell down and rolled about on the ground. We slid over an exquisite miniature mansion in red brick, with a charming pink pantile roof. The garden was beautifully laid out with flowers and shrubs, and it looked like something out of a fairy tale. Alerted by our dragon's roar, the owner and his wife came running out into the garden.

'What a beautiful house you have,' I called out to the lady.

'And what a beautiful balloon *you* have,' she answered.

By now it was time to land, for our fuel was running low. As usual when you want to land a balloon, you find all the fields for miles around are either full of barley or corn or flocks of hysterical cows or sheep, who would have a collective nervous breakdown if you landed amongst them, as would the farmers who own them. Finally, however, we spotted a meadow with no crop growing in it and devoid of domestic animals. To get to it, we had to drift over a large field of ripening barley and then over a belt of trees and then do a fairly smart three-point landing, for the meadow we were aiming for was narrow. It was at this point that the wind played us false. We had to come down low over the barley, skip skilfully over the belt of trees and then plummet down into the meadow. As we drifted across the barley, the wind suddenly faded and the balloon dropped earthwards with considerable speed. Jeff gave us a burst of flame to try to gain altitude, but it was no use; the basket crashed into the barley and then progressed across the field in a series of gigantic leaps, like a kangaroo. Three times we bounced, bone-breaking bumps, and then the wind caught the balloon and we were rushed across the barley, six inches from the ground, at breakneck speed, towards the menacing belt of trees, leaving a swathe of damaged crops behind us. The trees, prickly and dangerous, loomed nearer and nearer. Jeff did the only thing he could do – he pulled the cord that opened the flap and released the hot air from the balloon. Our big, bright, beautiful balloon shrivelled and died, but in its death throes the basket was thrown on its side precipitating us all in a heap on top of poor Chris. In its dying flurry the balloon dragged the basket a further fifty or so yards with all of us in a splendid tangle, trying our best not to sustain a broken leg or arm. At length, we came to rest and, bruised and breathless, crept out of the basket. The aluminium pole with the remote-control camera on it had been bent and twisted like a corkscrew, but fortunately the camera was not broken. What was more important, neither was any of us. Jonathan, Paula and the crew, who had been following our erratic flight in two cars, came pelting through the trees, looking extremely worried.

'Are you all right?' shouted Jonathan, obviously with visions of the talent with a couple of broken legs.

‘Yes, we’re OK,’ I shouted back. ‘It was as easy as falling off a log.’

Fortunately, they had brought with them the obligatory bottle of champagne which tradition demands that you consume to celebrate your first balloon flight. We drank it with relish, standing in the ruined barley field by the multicoloured corpse of our airy steed.

In spite of our rather frightening crash landing, we greatly looked forward to our flight on the following day, when we would be escorted by the helicopter. Unfortunately, in the morning the weather was unsuitable, but by mid-afternoon the forecast was favourable so once more we drifted aloft, the helicopter, with Chris hanging out of it, following us closely.

It was a glorious, golden afternoon, with the sky heat-hazy and a faded forget-me-not-blue. In this light, the countryside looked spectacular, the fields in so many different colours, emerald green, gold with buttercups, the pale fawn of ripening crops, the newly ploughed fields like reddish-brown corduroy. Once the helicopter signalled us on our walkie-talkie that Chris had obtained sufficient shots, we could let poor Jeff emerge from under the blanket and enjoy the flight. Lee, of course, had by now become so besotted with ballooning that she wanted me to go out the following day and purchase one. I confess that I was tempted, but I had to curb my enthusiasm and hers.

So, as the sun set and flooded the countryside with delicate greeny-gold light, we drifted on, quiet and haphazard as a dandelion clock, vowing that this was the only way to travel.

SHOOT EIGHT

OUR NEXT LOCATION was an extreme contrast. We left the rich lush green tapestry of the English countryside and went to the strange landscape of the Sonoran desert. To most people, the word 'desert' conjures up a harsh, waterless terrain of barren rock and sand, a place inimical to all life. This is, of course, true of some deserts but there are others in the world which are bizarre and wonderful places containing numerous plants and animals that have adapted wonderfully to this harsh environment. One of these extraordinary sections of the planet is the Sonoran Desert in the south-west of America, where you have thousands and thousands of square miles teeming with wildlife, studded with extraordinary forms of cactus and, in season, a myriad of gorgeous wild flowers. So, to show that a desert need not be as nasty as most people think, this is where we went to film.

Our crew consisted of Rodney Charters, affectionately known as Rodders: a stocky man who did everything at the run, even when burdened with heavy camera gear. He was always smiling, no matter what the difficulties, his eyes screwed up in a way that gave him an almost oriental appearance. His sidekick, Malcolm Cross, was a handsome young man with a luxuriant moustache and an air of being the sort of clean-living, clean-limbed young Englishman who made the Empire what it was. (It was Malcolm who wrote to me at the end of the shoot to say how much he had enjoyed it. He ended his letter by saying 'I came back in such good spirits that my wife is now pregnant.') Ian Hendry was our sound-man, with a wispy beard and soulful eyes that made him look like a middle-aged pixie. But in spite of his forlorn banished-from-fairyland look he took infinite pains with his job.

Our first day in the Sonoran Desert was a stunning experience. We had arrived at night and so could form no clear picture of what the desert was like, but at dawn the following morning we piled into the cars and drove out to visit the spots that Alastair had chosen as film-sites. To begin with, the sky was magnificent, a pale rose pink going to blood red where the sun was rising and flecked with lavender and yellow clouds. Against this, an army of giant Saguaro cactus stood silhouetted, like weird, spiky candelabra, some wearing crowns of ivory-white flowers with yellow centres. The Saguaro is probably the most spectacular cactus in the world, for it can grow to a height of fifty feet and they cluster together in forests that stretch for many miles. The cactus is mature when it is only seven or eight feet high, but at this point it is already fifty years old. From a distance, they look pleated, as though they were constructed from thick green corduroy. Along each of the pleats are bunches of stiff black spines some two inches long and as sharp as hypodermic needles. The whole growth process of this prickly giant is a slow one. It starts as a tiny seed and the first few years are precarious, for it has to contend with extremes of temperature from blazing heat to frost, from drought to floods. At this stage, it may be trodden on and killed by deer or partially eaten and stunted by rabbits or pack rats. If it can survive these hazards, then it grows slowly but surely. By the time it is seventy-five to a hundred years old, it is between twelve and twenty feet in height and then starts to develop its arms and its curious candelabra shape. The number and position of these arms vary so that no two Saguaro are alike. Some may have two arms, some twenty or as many as fifty. It is, like all cacti, a succulent, and like a huge prickly barrel it can store a vast quantity of water in its stem and arms. Its skin is thick and waxy, which of course makes it the perfect container for water. Its spines are not only a protection against the attack of animals such as deer or big-horn sheep, but also grow so thickly that they cast quite an appreciable amount of shadow on its trunk and arms, thus helping to keep the cactus cool in the intense heat. When a Saguaro dies, the flesh rots away and leaves a skeleton behind, a woody, basket-like structure that in life helped to support the barrel-like trunk and massive limbs. Inside these skeletons you can find odd wooden structures some ten to twelve inches or more in length that look like misshapen elongated Dutch clogs. These are in fact the remains of birds' nests. Because of its giant size, the interior of the cactus maintains a temperature several degrees cooler than the outside air and this makes it ideal for birds to nest in. It is the Gila woodpecker that, because it builds several nests each season, makes the Saguaro into a sort of prickly block of flats. Once the woodpecker has dug out a hole, the cactus (in self-defence) forms a hard, woody callus over the wound. These are the strange, misshapen 'clogs' that you find when the cactus dies. Once the woodpecker vacates its nest, other birds like owls, flycatchers and purple martins take them over, so it is possible to have three or four different species living in one of these cacti blocks of flats.

After we had driven some miles into the desert, we stopped and walked through the giant cactus forest. The Saguaro was the most prominent because of its size and impressive girth, but there were many other fantastic species as well. There were the Teddy Bear Chollas, for example, a medium-size cactus with many rather blunt limbs, so thickly covered with pale fawn-coloured spines that from a distance it looked like fur, and so the arms of the Cholla did look remarkably like the arms of a traditional teddy bear. Then there is the strange Boojum, with tall stems and long drooping arms, the whole thing covered with thorn-like black twigs so it looked as though each Boojum were in urgent need of a shave. These twigs bear leaves only when the Boojum has sufficient moisture to nourish them. These fantastic plants have been described as looking like upside-down carrots, though they are green not red. When you see some of them, sixty feet high with their drooping unshaven branches, it is really one of the oddest-looking plants of the desert. We were lucky that when we were there all the cacti were in flower, so that the desert was a riot of colour. There were flowers as green as jade, as yellow as daffodils, purple as heather, pink as cyclamen, tangerine orange and scarlet. If you had suddenly been dropped in the desert with this spiky profusion of strange shapes and the waxy, brilliant blooms and you had been informed you were on Mars, you would have unhesitatingly believed it.

Although the Sonora was hot, it was so dry that you did not really feel it. In fact, you had to be careful working out in the cactus forest, for you could get badly sunburnt without realizing it. An additional hazard of course was the cacti themselves, for they surrounded you on all sides with, as it were, their swords at the ready. Brush against a Teddy Bear Cholla, for example, and you soon found out how deceptive its furry, cuddly look was and you had to spend a tedious hour or so plucking spines out of your shirt or trousers.

Alastair, who always insisted on running everywhere and who was constantly tripping over his own feet, was in mortal peril most of the time we were in the Sonora. On one occasion, running backwards to get the right angle for a shot he wanted, he ran straight into an extremely prickly and unyielding Saguaro that had been growing in that spot for a hundred years or so and could see no reason why it should move for a film director. Alastair's hoot of agony could, with a following wind, have surely been heard in London.

We were very lucky to have the enthusiastic co-operation of the Sonora Desert Museum, a unique and wonderful institution that has living creatures for exhibition rather than stuffed ones. This of course enabled us to borrow most of our stars, and the majority of them were tame. But tameness can have its disadvantages, as we found out. We wanted to show the time-honoured method of catching lizards (which I have used with success all over the world) by the simple means of having a noose in a piece of fishing-line attached to a stick. You approach your lizard circumspectly, slide the noose gently over its head – a quick jerk and he is then yours. In order to demonstrate this technique, we borrowed one of the Desert Museum's oldest inhabitants, a large and venerable chuckwalla. These lizards, which are about two feet long, have fat, gingery-brown bodies, broad heads with a very Churchillian expression (only lacking the cigar) and extremely solid tails. The one we borrowed was called Joe and he gazed at us, plainly hostile, as though he had just finished making a speech of earth-shattering importance. We explained to him carefully what his part consisted of; we said that all he had to do was drape himself over a rock in the sand, wait for Lee to creep up behind him and slide the noose round his fat neck and, when he felt it tighten, he was to kick and struggle like a mad thing, as if he was a demented wild chuckwalla and not one that had been enjoying a privileged life for the last twenty-five years in the Sonora Desert Museum. From his highly intelligent expression we felt sure that he had understood his instructions and, since it was not a speaking part, he would be able to carry them out with aplomb. Alastair was convinced that here we had a star in the making and even went so far as to pat Joe on the head and murmur 'Nice snake' to him.

However, as soon as the cameras were set up and Lee, armed with her stick and fishing-line, was waiting in the wings, a strange change came over Joe. Draped on his rock, he ceased to be the agile chuckwalla we knew and loved. Overcome by what appeared to be a form of reptilian stagefright, he sat unmoving on the rock, looking like a splendid example of the taxidermist's art. Unblinking, unmoving, even when lifted off the rock by the noose round his neck, he looked as though he was stuffed to capacity with sawdust. Furthermore, nothing would break his trance-like state. We all shouted at him, waved things at him, threw delicious morsels like beetles in front of his nose, to no avail. Joe remained as immobile as if he had been carved out of rock. He was returned with ignominy to the Museum.

We had greater luck with the snakes. Steven Hale, who was our herpetologist guide and snake wrangler, arrived out at the desert location, the back of his truck full of wriggling bags of snakes, a sight which made the more faint-hearted of the crew recoil. The diamond-back rattlesnake he had brought was in a filthy mood and was rattling like volleys of musketry long before it was his turn to be emptied out of his bag to perform. He was a lovely snake, beautifully marked, and he rattled incessantly through his big scene and struck viciously at anything that came within range. A coral snake in pink, red, black and yellow like an excruciatingly gaudy Italian silk tie gave us some trouble because he had a turn of speed that was quite unprecedented and would disappear among the rocks in a twinkling of an eye. But probably the most handsome and certainly the most amenable was a five-foot-long king snake with jet-black, shiny scales, wonderfully marked in stripes of daffodil yellow. He had huge, liquid dark eyes and a most benign expression, for his mouth curved slightly, making him look as though he was smiling shyly at you. Placidly he allowed himself to be caught in Lee's noose, caught with a forked snake stick, to be discovered on rocks and under them, to slither endlessly through the cacti and other plant cover, to be handled endlessly, coiling lovingly round Lee's fingers, round her arms and round her neck. It was only finally, when Alastair said to Lee, 'Now put that lizard down on the rocks there,' that the snake, affronted, turned round and bit her. Fortunately king snakes are not poisonous.

One of the high spots of our desert filming, as far as I was concerned, was to see my all-time favourite bird in the wild – the road runner. With their wild eyes, their ridiculous unkempt crests and the loping run so reminiscent of all the lanky athletes you have ever seen, the road runner is the most comical and endearing of birds. We captured a curious incident on film which shows how, in the desert, nothing must be wasted in this harsh environment. There was a nest with three baby road runners in it, and one of the chicks had died. To our astonishment, when the mother discovered this, she picked up the dead baby out of the nest and proceeded to feed it to one of the other chicks. When we last saw it, the chick had succeeded in engulfing the head and neck of its dead brother, while the body dangled outside. This, apparently, is quite usual with road runners, for they will sometimes catch and kill snakes that are too large for them to swallow in one go, so they swallow as much as they are able to and leave the rest dangling outside. When half the snake has been digested, they can swallow the other half.

It was while we were shooting in the desert that we had

one of those awful days that make filming so unpredictable and so irritating for everyone concerned. In an effort to show every aspect of desert conditions, we had filmed cactus desert, scrub desert, stony desert and grassland desert. All that remained was to film what most people consider to be typical desert – mile upon mile of rolling sand dunes. Alastair, during his reconnaissance, had found the ideal spot some fifty miles away. Here, three- and four-hundred-foot dunes, beautifully sculpted by the wind and rain, stretched in every direction as far as the eye could see. Moreover, a main highway ran right through the area, thus making access easy. Alastair waxed so lyrical about these dunes that I got the very strong impression that they would make Outer Mongolia, the Gobi and Sahara deserts pale into insignificance. So, thoroughly overexcited at the prospect of shooting scenes that would rival, if not indeed surpass, anything that Hollywood had depicted in *Beau Geste*, we got up very early and drove off into a dawn that was a blur of golden light with tiny clouds like feathers picked out in scarlet and purple.

Alastair had been to see the dunes fifty miles away in California on a weekday, when their silent majesty had so impressed him. This was a Sunday, and after driving for several hours we arrived at the dunes to find a very different set-up from the one Alastair had depicted. True, there were huge, beautifully sculpted dunes; true, they stretched for miles in every direction; true, they looked like a Hollywood desert, so that you expected to see Ramon Navarro gallop over the horizon at any minute. However, instead of the Hollywood heart-throb, what you did have was what appeared to be three-quarters of the Californian population and smelly and extremely noisy dune-buggies ripping the hell out of the dunes. There were hundreds of them, skidding, bouncing, roaring, screaming, making any idea of doing a sound-take impossible. Indeed, one could hardly hear oneself talk, quite apart from the added distraction of having half a dozen dune-buggies waltz across the sand past you, inhabited by a vast selection of scantily clad and very nubile-looking young ladies. In desperation, we drove on and on, hoping to find a section of the dunes uninhabited, but the whole area buzzed with dune-buggies like a disturbed wasps' nest. Eventually, in despair, Alastair suggested that we turn round and head back to the beginning of the dunes (which seemed somewhat less inhabited) and content ourselves with some silent filming. Rodney, who was driving, and who had a fine disregard for the laws of the land, did a U-turn in the centre of the highway and headed back the way we had come. Within seconds (or so it seemed) a Big Brother helicopter (bristling with police) had relayed the enormity of our crime to a police car below, and with blaring sirens we were overtaken by it and flagged down. The dark-uniformed policeman who gave us our ticket looked formidable. Quite apart from his gun (which you knew instinctively he could use to blow the ace out of a playing-card at four thousand feet) he was built, roughly speaking, on the lines of Mount Everest and was not only obviously adept at boxing, karate and ju-jitsu, but could probably fly better than Superman if you provoked him. The whole air of menace was accentuated by the fact that he was so cordial and soft-voiced. Even Alastair, who at the best times was no respecter of authority, was intimidated by this soft-spoken hunk of manhood, who looked as though he ran the whole of the CIA singlehanded. We took our ticket meekly.

We drove on until we reached the place where the dunes stopped, and here Alastair pointed out that at one side of the road dune-buggies were prevalent but on the other side they were non-existent. So, acting on our director's instructions, we drove off the highway down a rough track that led through the dunes. It was at this point that we discovered why this area was free of dune-buggies. At a section of the road most distant from the highway and civilization, our car sank in over its axles and got stuck. Paula, Lee and I had to walk two miles to the highway and then another two miles before we found the garage that had the necessary truck to drag us out of the dune we were embedded in. We got back to our hotel very late that night, feeling exceedingly frustrated and irritated by the fact that, in addition to not having been able to shoot any film, we owed the Police Department twenty-five dollars in fines.

But this was our only bad day; the rest of the shoot in the desert was perfect. The weather was flawless, the temperature superb, from dawn with its wonderful green, pink and lavender clouds fading to crisp sunlight that enveloped the cacti in a blurred golden web of light, to the evening when the huge sky (the sky looks twice as big, somehow, in the desert) was drenched in scarlet and purples of such brilliance that they would have made a Turner sunset look anaemic in comparison.

The fascinating thing about shooting this series was the contrast. One minute you would be filming in snow and the next minute sweating in the heat of a tropical forest. One minute paddling a canoe down an English river, the next minute paddling a canoe over a tropical reef. So in this case we had a contrast, for we left the giant cactus forests of Arizona and flew down to the rolling grasslands of southern Africa, to that great game reserve with the marvellous name – straight out of Rider Haggard – of Umfolozi.

Approaching the wonderful reserve is one of the most salutary and frightening biological eye-openers I have ever experienced anywhere in the world. You drive through mile after mile of rolling green grassland that reminds you vaguely of parts of England. You are also vaguely aware that forests must have been felled to create this grassland and you are aware that, while it looks superficially lush and green, it is in fact desiccated and eroded, overgrazed and overpopulated. However, this does not really impinge upon you until you reach Umfolozi. You are driving through these rolling green hills, eroded and sparse, and then suddenly you see ahead of you a fence and beyond that fence is what Africa was like before the advent of the white man and before the Africans had overpopulated. Wonderful rolling acacia scrubland, rich meadows, giant pot-bellied baobab – a rich lushness that had to be seen to be believed.

Those of my readers who, like me, are tottering on the borders of decrepitude may remember Judy Garland in the film called *The Wizard of Oz*. They will recall how her house is whirled up over the rainbow by a tornado. Up to that point, the film had been in black and white, but when the house crashes to a standstill and July Garland timidly opens the door everything is in Technicolor of the most flamboyant sort. Arriving at Umfolozi had very much the same effect on me. We had been travelling through a man-made and man-desecrated landscape, but you were not fully aware of what your species had destroyed because there was no contrast in Technicolor, as it were. But arrive at the fence that guards this chunk of original Africa and even someone like myself (who is fairly aware of the world's problems of conservation) is jolted. You suddenly realize that you have been driving through a man-made equivalent of a desert and have arrived at an oasis – behind bars.

As you enter the park, not only do you have the extreme contrast of vegetation but suddenly the landscape is alive with animals. Zebras, striped like Victorian humbugs, cantered alongside the truck, throwing up their heels skittishly. With them cavorted the brindled wildebeests, or gnus, their curiously twisted horns making them look as though they were peering at you over a pair of spectacles. For such ungainly animals they are astonishingly agile. A herd of gnu taking off is more like a ballet than anything, for they twist and buck and prance, one minute practically standing on their heads and the next minute leaping into the air and executing a complex pirouette. As the zebra and gnu galloped through the undergrowth, they disturbed flocks of plum-purple starlings and groups of ground hornbills with huge curved Fagin-like beaks and scarlet wattles. They paced along as sedately as soldiers on sentry duty and gazed at us out of huge, soulful eyes, framed by thick, extremely sexy eyelashes. We had travelled about a mile through the park when we saw its most important inhabitants – a white, or square-lipped, rhinoceros. These huge and magnificent beasts (the largest land mammal next to the elephant) were at one time driven to the edge of extinction. Fortunately, at the eleventh hour, action was taken to preserve this antediluvian giant and so now in Umfolozi and other parts of South Africa they are on the increase. This was a huge male and he moved majestically through the trees, his enormous head with a four-foot horn curved like a scimitar on his square-lipped nose. Several tick birds perched along his back, like ornaments on a mantelpiece. Occasionally, as the rhino's massive legs brushed through the grass, they would disturb small animals or grasshoppers and the tick birds would fly off their moving perch, catch an insect and then return to the rhino's back to eat it. We stopped the car within thirty feet of him, and he came to a halt and surveyed us calmly. Then, uttering a deep sigh, he crossed the road in front of us and disappeared among the acacias.

Not more than half a mile further on, we came upon a group of what I consider to be one of the most beautiful of all mammals, the giraffe. There were five of them; three were quietly browsing on the acacia-tops while the other two, who were obviously on honeymoon, were behaving in the most ludicrously besotted manner. Facing each other, they were managing to entwine their necks in the most astonishing manner, more as though they were swans than giraffes. They were kissing each other rapturously, their long tongues sliding in and out of each other's mouth voluptuously, with the sort of passion you expect from a French film, but somehow don't associate with giraffes. They, like all lovers, were completely oblivious of everything except each other and they took no notice, even when we got out of the car and walked quite close to them. Eventually, we arrived at the singularly unattractive series of cement-block buildings constructed by the South African government to make the tourist feel loved and wanted. It was rather like living in a badly designed public lavatory but was more than compensated for by our surroundings.

Our cameraman here was another Rodney – Rodney Borland, and his wife Moira. Together they had made some superb wildlife films and so knew the African bush intimately.

It was at this time that Alastair was having a prolonged and intense love affair with a mole. Perhaps such a statement needs some elaboration. I had proclaimed that I had no intention of going to South Africa unless I was allowed to meet a creature that had long fascinated me – the golden mole. There are several species of this strange beast and, although bearing a strong resemblance to the European mole, they differ from it chiefly in having fur that is extremely silky and that glows like spun gold. Bearing my wishes in mind, Alastair had gone to considerable trouble to get someone in Durban to extract a golden mole from his garden and hand it over to us for the filming. It was an enchanting creature, with eyes so minuscule that it looked like somebody who had mislaid his spectacles. It was about five inches long and looked like a furry ingot, scuttling about in its box of earth. Like most insectivores, of course, it had a voracious and insatiable appetite and required about three hundred yards of worms per day to keep it cheerful. For some reason, Alastair worked up a great affinity with this curious little creature, digging up vast quantities of worms for its breakfast, lunch and dinner and keeping it in his room at night. He did admit that, as McTavish spent the entire night trying to dig his way out of his box, he had known more restful sleeping companions. Although the golden mole bears a superficial resemblance to the European mole, they are not related and the likeness has come about simply because they have both adapted similarly to a fossorial way of life and thus have developed similarities, such as the powerful forefeet for digging, vestigial eyes and strong, bulldozing snouts. McTavish, as I say, looked golden in most lights, but if the sunlight struck his glossy fur at a certain angle he could turn green, violet or purple – a really striking display for a mammal. One night, McTavish's nocturnal activities were successful. He found a weak spot in his box and with his powerful forefeet enlarged this to a hole. Alastair, heartbroken, reported at breakfast that he had awoken to find himself moleless. It was fortunate that we had shot all of McTavish's vital scenes before he made his bid for freedom.

One of the things that we wanted to show was how the various ungulates in savannah lands have each developed their particular grazing habits so that, for example, a giraffe will graze off the tops of the acacias, whereas the kudu will browse lower down the tree. In this way, by splitting up the various levels of grazing, there is less competition and the food supply is more evenly shared. We thought that the best way to demonstrate this would be to show two extremes – a creature that browsed at the very tops of the trees and one that fed at ground level. So we decided on a giraffe and a tortoise as our examples.

First, with considerable difficulty, we succeeded in finding a large, somnolent tortoise sitting under a baobab tree. Alastair, who had been getting increasingly jittery as no tortoise seemed to be forthcoming, sighted this reptile, who appeared to be in a trance, and he leapt from the car with cries of joy and swept the tortoise into his arms and clasped it to his bosom. Now, this is not the wisest thing to do to a tortoise when it is fully alert. To do it to one that appears to be quietly sitting under a baobab reciting one of the longer and more boring of Tennyson's poems to himself is courting disaster. All tortoises have large and retentive bladders, and this one was no exception. To say that Alastair was drenched would be an understatement. He was most upset.

'Nobody told me that tortoises peed,' he kept saying plaintively. 'Nobody told me that they were so copious.'

So we put the now empty tortoise into a box, dried Alastair as best we could and then set off in search of a giraffe. As always happens, of course, there was not a giraffe to be seen. Normally, the landscape was littered with them, but now we could not track down a single one. After driving around for several hours, however, we finally found one – a gigantic and beautifully marked male, lurking among the acacias.

Alastair's great idea was that I should approach the giraffe, taking the tortoise with me. When I got as close to the giraffe as he would let me, I was to put the tortoise on the ground, face the camera and start talking about the browsing habits of giraffe at the tops of the trees and then point out that lower down other antelopes grazed and then right at the bottom you got a grazing creature such as a tortoise. At this point, I was to bend down and pick up the tortoise. Like most of Alastair's ideas, this was easier said than done.

I got out of the car and, bearing the indignantly hissing tortoise in my arms, I approached the giraffe. He watched my approach with an expression of complete incredulity on his face. During his long and happy life, fate had never engineered it that his lunch would be interrupted by a human being carrying a vociferous tortoise, and he was not at all sure that this was an experience he wanted. He gave a tremulous snort of alarm and walked round to the other side of the acacia so that only his head was visible.

'That's no good,' hissed Alastair. 'I want his whole body.'

Slowly, I followed the giraffe round the acacia and slowly he moved round it, keeping a large section of prickly tree between me and him. I continued to follow him and for quite some time we went round and round the tree, as if we were doing an old-fashioned waltz.

'It's no good,' I said to Alastair, 'you'll have to move the damn camera.'

So the camera was moved and after a lot more waltzing round the acacias I finally got the giraffe into the position that my director required.

'Excellent,' said Alastair excitedly. 'Now put that snake thing on the ground and talk about zebra.'

So I put the tortoise on the ground, straightened up and spoke long and eloquently about the giraffe and their feeding habits and about the feeding habits of other ungulates.

'And so,' I concluded, 'by grazing in this selective manner the food is evenly distributed from the very tops of the trees to the ground level, where you get grazing animals like this.'

I bent down to pick up the tortoise and, to my astonishment, there was no tortoise there. In a burst of speed unprecedented in such a reptile, he had fled and was fifty feet away, making for the peace and tranquillity of the acacia groves. Needless to say, this whole sequence was not a success.

Another of Alastair's brilliant ideas was to have me start the programme and set the scene while standing, as it were, hand in hand with a white rhino. So besotted had he become with this idea that we spent three days doing nothing else but driving round looking for white rhinos. We had no difficulty in finding them, since the park was overflowing with them. The difficulty lay in trying to get them to co-operate with Alastair. We found a portly mother and her plump child sitting about in a waterhole that they were companionably sharing with a buffalo. The buffalo had mud all over his back and shoulders which had dried and cracked, so that he looked as if he were wearing a grey jigsaw puzzle. The female rhino and her baby were not aware of our presence and it was possible that I could have got close enough to complete the scene to Alastair's satisfaction if it had not been for the buffalo. He had been standing belly deep in the waterhole, sunk into that bemused state that overcomes all buffalo when they get anywhere near water, and so he woke up with a start when he suddenly saw me getting out of the car. By this time, his massive weight had made him sink so deeply into the mud that, when he tried to vacate the waterhole, his legs stuck and he fell sideways, thrashing about wildly. The rhinos, not unnaturally, took this to be a sign that something was amiss and so, as the buffalo finally righted himself, they all left the waterhole at a brisk run and disappeared into the trees. This sort of thing happened time and again. Rhinos, being shortsighted, make up for this defect by having extremely keen hearing and a good sense of smell. Also they are exceedingly suspicious, probably because of their bad eyesight, although what enemies a creature of such massive proportions could have was a mystery to me. However, all our attempts at getting me and a rhino standing side by side were meeting with failure and it looked as though we were going to have to leave South Africa without this vital opening shot that our director insisted upon.

It was our last morning and, amid groans of despair from all of us, Alastair insisted on driving out to have one final attempt to get me and a rhino together. It was very early in the morning and for this reason, I think, we were successful, for the massive old male we finally found looked very bemused, as if he had just that minute got out of bed. Cautiously, we drove the car over the savannah towards him, keeping downwind. When we were within some forty feet of him, we switched off the engine and discussed the situation in hushed whispers, whilst the gigantic creature stood there, flicking his ears to and fro suspiciously. He was dimly aware that something untoward was going on, but was not sure exactly what. Another point in our favour was that he had no tick birds on him, for these invariably gave the alarm that sent their steed lumbering off.

'Now,' whispered Alastair, 'what I want you to do is to get out of the car, get as close to him as you can, then turn round, face camera and do your opening speech.'

'Thanks,' I said. 'You, meanwhile, will be skulking in the safety of the car.'

'I shall be with you in spirit,' said Alastair.

So, hoping all the stories I had read about the ease with which you can dodge rhinos because of their bad eyesight were true, I got out of the car. When I had vacated the safety of the car and was walking towards it, the animal, for some reason or another, appeared to grow to twice normal size. Slowly, holding my breath and attempting not to tread on any dry twigs, I crept up towards him. He lowered his gigantic head, snorted and flapped his ears to and fro with a noise like someone cracking a whip, which was not the most reassuring noise that I had ever heard. His horn appeared to be twice as tall as the Eiffel Tower and much more pointed. I got to within about twenty feet of him, which I reckoned was as close as I wanted to get. Then, taking a deep breath, I turned my back on him stalwartly and, beaming at the camera and trying not to show how scared I was, I began my opening to the programme. I was halfway through it when I heard a rustling, thumping sound behind me, which took several years off my life. Any minute, I expected to be lifted off the ground on the tip of that scimitar-like horn. I took a hasty glance over my shoulder, trying to look casual, and found to my infinite relief that the rhino had swung round and was moving away from me, snorting irritably to himself. I turned with relief to the camera and finished my introduction without, I flatter myself, a tremor in my voice, but turning my back on two thousand pounds of rhinoceros was one of the hardest things I had to do in South Africa.

SHOOT NINE

SO FROM THE AFRICAN SAVANNAH, we flew back to springtime in England. Early Spring in England can be beautiful with pale-blue skies, banks garlanded with butter-yellow primroses, woods full of the mysterious woodsmoke-blue of the bluebells, fields golden with buttercups and kingcups, the first delicate shimmer of green buds on the trees, the sun frail but warm.

Not, however, if you are making a film.

Our next programme was to be about an English pond and an English river, both fascinating, especially in spring when all the aquatic creatures from toads to newts, from otters to mayfly, were starting breeding activities.

Not this spring.

This was a film-maker's spring, with leaden skies, arctic temperatures accompanied by rain, hail, sleet and finally, when we thought the weather had run its gamut, snow. The beautiful pond which Jonathan had picked out for its translucent, amber-coloured waters like fine pale sherry turned into a muddy, opaque broth in which nothing could be seen. The River Wye, our second location, which normally rushed joyfully over its rocky bed, clear as molten glass, was transformed by mud and debris into something closely resembling one of the more unpleasant lava flows regurgitated from a bad-tempered and dyspeptic volcano. Needless to say, this had a distressing effect on Jonathan, and every time he looked out of the window he – mentally, as it were – fell on his sword. We rushed from one location to the other (at opposite ends of the country, naturally) in the hopes that the weather had cleared, but in vain. Paula was in despair for it was her job as producer to keep everyone's spirits up, but climatically speaking this was impossible. In addition, during the course of the various shoots, she and Jonathan had very unwisely fallen deeply in love and had decided to get married when the series was finished, so as a prelude to normal married life what could be more natural than that Jonathan should attribute the prevailing inclement weather conditions to his betrothed. It was a trying time for us all.

'Look, honey,' she said very sensibly, 'why don't we go and film Lee shooting the rapids? In those shots it doesn't matter how muddy the water is.'

'What a good idea,' said Lee, who was dying to have her first try at white-water canoeing. 'Let's do that, Jonathan.'

'It might help you to feel better if you risk my wife's life in the rapids, sadist that you are,' I pointed out.

'Yes,' said Jonathan gloomily, 'I suppose we could do that.'

So we packed up and drove away from the mud-coloured pond to where the River Wye rushes and twists through black rocks. Here the dark muscles of water curved round the rocks in great bursts of foam and everything was a roar and chatter of water. Lee, thoroughly excited, was decked out in a scarlet wetsuit and a becoming bright-yellow crash-helmet. Then she was wedged into a slender and fragile-looking canoe and launched into a placid area of the river for her first and only lesson. Such is the perversity of women that within half an hour she was handling the canoe in a manner equally (or more) professional than her instructor.

The purpose of all this was to show how a canoeist had to use the strength of the river to his or her advantage, using the force of the water as propulsion, using the current to steer and the curving eddies as placid areas of calm, like water parking-lots. We were then to illustrate it with the creatures that live in these turbulent waters, using exactly the same methods for their survival. So the camera was stationed on the rocks by the white water and Lee, wedged in her canoe, waited a quarter of a mile upstream for the signal to shoot her first rapids. Attached to the bows of the canoe was a tiny camera, and a cord ran along the edge of the craft to a button near where Lee was sitting. The idea was that as she reached white water she was to press the button and then take a close-up film of herself and the canoe zooming through the waves, water splashing over her. Meanwhile, the other camera on the bank would be getting all the wide shots. So her canoe set off, skimming between jagged black rocks, bucking and bouncing on the shining streams of water, the nose of the fragile craft digging into the bursts of foam like a pig searching for truffles in a bouquet of white roses. I must say Lee handled the craft as though she had been doing it all her life, but I watched in some trepidation and heaved a sigh of relief when she could turn into a placid area of water and bring her canoe to a halt. It was then that we discovered that, although she had switched on the little camera in the bows, during her vigorous efforts to avoid being overturned by a rock she must have knocked the button again with the paddle, thus switching the camera off. There was nothing for it but to do the whole thing all over again. So the canoe was carried a quarter of a mile upstream and my wife (by now, of course, considering herself an old sea-dog) incarcerated in it, and once more she shot the rapids, the canoe sliding and leaping like a spawning salmon; and this time, fortunately, the camera worked.

It is strange to think that all the great rivers of the world, the Amazon, the Nile, the Mississippi, share the same humble beginnings – a few teacupfuls of water bubbling out of the ground – then as the water hurries down to the sea it gathers momentum and force. It changes from a tiny skein of water into a broad, majestic river. Rivers, whether large or small, are the veins and arteries of the land, and along their glittering lengths they give home and food to a vast band of creatures that live in, on or alongside them.

One can understand a host of creatures living in the placid world of a pond but it is more difficult to reconcile yourself to the fact that many creatures have adapted themselves to the more turbulent areas of a river. We had already got film of some of the more extraordinary of these, filmed naturally under controlled conditions, to enable us to get big close-ups of the way they manage to survive in this boiling turmoil of water. Take the common caddis-fly larvae, for example. In any placid pond, you may find those who have spun themselves a silken tube to live in and then camouflaged the outside with sand or tiny bits of vegetable debris. (When I was young, I used somewhat unfairly to remove a larva from its home and then, while it was spinning another, provide it with different-coloured materials, such as brick dust and powdered slate, and thus get multicoloured caddis larvae.)

In the still waters of a pond a camouflage of plant debris will suffice, but in a fast-flowing stream or river the larva needs something more substantial to help anchor it and prevent it from being swept away by the current, so it uses tiny pebbles which to the larva are as big as boulders. When you are examining the pebbly bottom of a stream you are sometimes taken aback when what appears to be a small pile of pebbles walks away. Another species of caddis does not build itself a silken caravan but has another method of coping with the current and turning it to its own advantage. It chooses a cave between the pebbles as its home and then over the mouth of this cave it spins a net and sticks stones to the edge of this trap to prevent it from being swept away. Then, like a Victorian spinster lurking behind a lace curtain, it waits patiently for the bountiful river to fill its net with food.

Another creature that, though so small and fragile, copes wonderfully with its violent environment (which must be for these little creatures the equivalent of us living under the torrent of Niagara) is the black-fly larva. This curious creature looks like a tiny, elongated caterpillar, with a huge pair of Edwardian walrus moustaches at its head. This creature makes a sort of pin-cushion of mucous on the rocks and then attaches itself to this with a series of sharp hooks at the end of its body. Then it stands up on this cushion and sifts particles of food from the river as it sweeps past. It is a curious sight to see a creature that is apparently feeding itself with its own moustache. Another astonishing beast of this sort of environment is the side-swimming shrimp. It looks not unlike one of the so-called sand-fleas found so commonly on beaches, but one that has been run over by a steamroller and flattened, so it is forced to swim on its side, but in fact this slim, flattened body gives the least resistance to the current and its shape enables it to dart from one slender crevice to another in the stream-bed, wedging itself in firmly so that it avoids being twitched downstream by the thrust of the water.

At length, in spite of the weather, Jonathan was reasonably satisfied. As well as all the necessary linking shots of Lee and myself, he now had film of beautifully marked polecats and those excellent swimmers, the water voles, the elegant mink and a family of baby coots who, in their yellow down jerseys had scarlet faces which made them look as though they were suffering from chronic high blood-pressure. Curiously, they resemble so many of the punks that you see about, but they are infinitely more appealing. We also had a lovely sequence of swans, those avian giraffes of the river, who sail majestically along, delving deep to reach the weed and then tossing it elegantly over their shoulders to be feasted on by the flotilla of fluffy grey young swimming expectantly behind them.

So we left the river and went back to the pond which, though still muddy, was not quite as disgusting as it had been. Here we had several sequences to do, one involving a boat and a piece of film showing me walking on the water. Ponds are like little worlds of their own, and an enormous variety of creatures depend on them. Unfortunately, all over the country the number of ponds is dwindling as they are drained and filled, being considered by the nature-loving British farmer as being useless bits of water that get in the way of crops or grazing. The fact that an enormous number of creatures, from frogs and toads and dragonflies to the myriad microscopic beasts, depend on these ponds for their very lives does not concern a well-educated modern society.

Fortunately, some people do still care and are willing to help and not harass nature out of existence. There is, for example, the excellent Frog Watch conducted by the Royal Society for Nature Conservation. In Britain you can phone a frog – that is to say, there is a special hot-line telephone, number broadcast on local radio and printed in the local paper, by which you can report your first sighting of frog or toad spawn in ditches, garden ponds, or the dwindling number of natural ponds. Experts then mark this on large-scale maps of the area and thus gain a picture of the extent of the amphibians' breeding-grounds. Frogs, of course, stick fairly close to their birthplace throughout their lives, but toads present a very difficult problem. As soon as the toadlets come out of the pond they spread far and wide, for their skins do not require the moist environment that the frogs need. However, when they grow up and the breeding season arrives they hop off in their thousands to the pond or lake where they were born. Of course, in many cases they have to cross roads or even motorways to attain their objective and so thousands are killed annually by cars. In the Netherlands, where they seem to deal more sympathetically with their wildlife, they have created underpasses for migrating toads. There is as yet no such refinement in Britain, but there is a move afoot to remedy this with the slogan 'Help a Toad Cross the Road'. People empathetic to toads (and who could be otherwise since, if kissed, each one is a potential prince?) take buckets, dustbins or other containers to those points where the toads habitually cross and as the vast concourse of amphibians arrives they bundle them into the buckets and other containers and take them safely across the road. It is to be hoped that the Boy Scouts give up their time-consuming traditional task of helping old ladies across the road and concentrate instead on the toads.

Of course, we had already filmed quite a number of the pond sequences under controlled conditions and had got some remarkable material. There was, for example, the curious little fish called the bitterling that uses the fresh-water mussel as a sort of babysitter. In the breeding season, the female bitterling develops an extraordinary long, white, slightly curved ovipositor, which looks as though it has been made out of white plastic; then, accompanied by her husband, she goes in search of her babysitter. The fresh-water mussels, about four or five inches long, lie on their sides in the mud and look very like oval, slightly flattened stones. At one end of the shell there are two siphons – one exhalant and one inhalant. The mussel sucks in water through the latter, extracts what food it contains and then expels the water now filtered of its nutrients out of the other siphon. Both these siphons look like little round mouths and are capable of being shut tight by the shell should it become alarmed. The bitterlings seem to realize this and so, having decided which mussel is to become their nanny, they gather round it and repeatedly bump it with their heads. This of course panics the poor bivalve and both siphons are firmly shut against this potential danger. However, the bitterlings continue their battering-ram activities and eventually the mussel decides that this constant knocking on its shell must be harmless so it relaxes and the siphons open and start to act normally. This is what the fish have been waiting for. The female swims over the mussel, stabs her long ovipositor into the exhalant siphon and lays her eggs, which are like small white ping-pong balls. (Until this process was filmed for this programme, it was always assumed that the inhalant siphon was the one used by the fish.) As soon as the egg is laid, the male moves over the siphon and fertilizes it. Occasionally, in jerking out her ovipositor the female pulls the egg out as well. This is promptly eaten by one or other of the parents. 'Waste not, want not' is a law and not just an adage in nature. When the batch of eggs has been installed in the mussel and fertilized, the parents forget about them and the rest of the process is left to the babysitter. It is a very curious process, for what happens now is quite startling. When the bitterling eggs hatch, the mussels spawn and from their eggs come the future generations of mussels, which look at this stage like tiny castanets with hooks on them. These fasten themselves with these hooks on the baby bitterlings and as the young fish leave their shell nursery and swim away into the pond they carry with them a host of baby mussels which eventually drop off to form other mussel-beds far away from their parents.

We also filmed the activities of an extraordinary spider. If asked to look for a spider, the last place one might suggest is the bottom of a pond, yet it is here that the extraordinary water spider builds its home. Between fronds of weed, it constructs what is to all intents and purposes a diving bell, a silken bowl upside down, which the spider then fills with bubbles of air carried down to its home on its hairy legs. Around the bowl it spins a web as ordinary spiders do and then it lurks in its submarine home until a tadpole or a waterboatman or some such creature blunders into it, whereupon it rushes out and captures its prey. An early naturalist had described how the spider, when the air in its home grows stale, removes it and replaces it with fresh; but, as this had only been observed once, people thought that maybe the naturalist was mistaken, but we actually filmed this curious piece of behaviour. The spider comes to the top of the bell, tears a small rent in the silk, allowing a bubble of air to escape which it catches with its legs and takes to the surface to release; and on its return journey it brings with it a fresh bubble of air to freshen the bell, exactly like a hostess emptying ashtrays and opening windows after a cocktail party.

Possibly some of the most fascinating of the pond denizens we got on film were the planarians. The species we filmed were strange, éclair-shaped creatures that as they glide as smoothly as quicksilver about the mud look as if they were manufactured out of damp black velvet. They are of course flatworms and look vaguely like aquatic slugs. They are hermaphrodite, each animal having both male and female organs and producing both eggs and sperm. However, the eggs from one planarian must be fertilized by the sperms from another. They feed principally on dead matter such as tadpoles or tiny fish, tearing and sucking at the meat and the juices of their prey. They can, however, exist for very long periods without food, but then they gradually get smaller and smaller, since they are literally eating themselves. Another unusual aspect of these curious little creatures is that the mouth is used both to ingest food and to excrete. Their reproduction sounds like something out of science fiction, for not only do they lay eggs but, should one be cut in two by accident, two new planarians grow from the two halves. In some species they normally increase their numbers by having a sort of tug-of-war with themselves, tearing themselves in two to swell the population. There was a fascinating series of experiments conducted on a species of American planarian which proved that they could be taught with the aid of weak electric shocks to select either a black or a white tube as the correct escape route from a maze. Moreover, if the planarian was cut in two both bits could remember this lesson. Even more amazing, it appears (though this has not been thoroughly investigated) that, if a trained planarian is devoured by an untrained one, the untrained one 'inherits' the trained one's knowledge. If this is true, it is surely one of the most remarkable pieces of animal behaviour. It is as though a schoolboy ate – suitably roasted – his schoolmaster, and thus obtained his knowledge and experience. It is reminiscent, of course, of the human belief that if you ate your vanquished opponent after battle you would inherit his courage and strength.

We now came to the two sequences, one involving a boat and the other what were laughingly called watershoes, which we needed for the walking-on-the-water sequence. The shoes were a very strange contraption. If you can imagine two six-foot-long slender canoes, joined together by jointed rods, and each canoe ending in what looked like half a dolphin's tail in rubber or plastic, you have some idea of this curious means of progression. The way you used them was this: you put one foot into each canoe by sticking it into the canvas top, and then you seized hold of the rudder, a long pole that ran to the bows of your craft, then with somebody's assistance you were launched. As soon as you were afloat, you stamped your feet up and down as if marking time in one spot. This movement had the most astonishing effect on the two halves of the dolphin's tail, making each piece flap up and down, thus propelling you through the water. It was quite an exhausting and laborious business and you discovered whole sets of muscles in your legs that you were unaware ever existed. The dangerous part of the business was that if you lost your balance and fell over it was extremely hard to extract your legs from the canvas covering the top of the shoes and so you stood a pretty good chance of drowning before anyone came to your rescue.

The rowing boat Jonathan had found was a magnificent craft some ten feet long, broad in the beam, so that she resembled a fat beetle in shape with the paint peeling off her in great strips like skin off an unwary sunbather. So, while I stamped about the surface of the pond in my water-shoes, Jonathan rowed the boat with the camera crew in it after me. When I had done all the walking-on-the-water shots to his satisfaction our noble sound-recordist, Brian – who had been watching my performance enviously – was determined to try his hand at it. We launched him successfully, and in great style he completed a circuit of the pond. It was when he was coming in to land that he ran into trouble. For some reason, as he reached the shallows, he lost his balance, fell over sideways and lay there in two feet of water, his legs trapped in the watershoes, frantically struggling to keep his head above water. It was fortunate that the water was shallow so he could keep one hand on the bottom and so keep his head above the surface. If the water had been deeper and we had not been handy as rescuers, he would assuredly have drowned.

The next sequence we had to do was Lee and I rowing the boat about while I explained how an amateur naturalist does not have to spend large sums of money on vast quantities of sophisticated equipment but can, with a little bit of ingenuity, convert everyday appliances to his or her needs, such as making a perfectly good grapnel out of a wire coathanger, which will enable you to pull patches of weed from the centre of a pond to the bank. As every amateur naturalist knows, all the most exciting bits of weed are invariably situated in the exact centre of any body of water and so require some sort of appliance to bring them within reach. To get these idyllic shots of Lee and myself, clad in straw hats, rowing gently across the placid pond was slightly more complex than one imagined. To begin with, the boat was not very large and so once Lee and I were in it there was very little space left in the stern for anyone else. Finally, with the boat sagging low in the water, there was the cameraman, the assistant cameraman, the sound-recordist and the director, as well as Lee and myself, in our long-suffering craft and my long-suffering wife had to row this crew to and fro over the pond until Jonathan was satisfied that he had got the shots that he required.

So we left the turgid pond and the damp, cold English countryside and sped across the Atlantic to what Americans for some reason best known to themselves call the Big Apple – New York City. Here, with Paula watching over us, Alastair directing and Rodders doing the camera work, we were going to try to show that to a dedicated amateur naturalist even a gigantic modern city could provide grist to his mill. Alastair greeted us as only Alastair can – broad grin, glittering eyes, head on one side because that crafty hangman has placed the knot snugly under the left ear.

'Worms,' he said by way of greeting, 'worms, beating the ground to feel like rain . . . cemetery . . . lots of life in a cemetery.'

I tried to think of all the cemeteries I had been in – austere white ones like hospital wards, lichen-encrusted ones where you had to fight your way with a machete to decipher the carunculated faces of the gravestones with their time-blurred messages. I have never actually flipped back the lid of each grave to display life. The whole idea of finding life in a cemetery was a novel one and worthy of Alastair's at times macabre sense of humour, so we went to Calvary cemetery.

As a cemetery, it was quite extraordinary: it contained not only ordinary gravestones such as are used for plebeian folk, but monstrous mini-mausoleums looking like crosses and clones of the Acropolis and St Paul's Cathedral, generally, as far as I could judge, sheltering the mortal remains of somebody called Luigi Vermicelli or Guido Parmesan. I think probably the most horrifying thing about this was that it lay on a gentle slope of land, each monument to the dead as white and pure as any ski slope or emergent mushroom you have ever seen, and looking down through the strange vista of monuments to the dead (which made you feel that God must have been a confectioner of some skill) your eyes were rewarded with the New York City skyline mirroring and enlarging the graveyard at your feet. You could be pardoned for asking where the skyscrapers ended and the graves began. In fact, you began to wonder if the skyscrapers were not enormous mausoleums and was it really worth while to use up so much useful land by bringing out your dead. However, I was proved wrong, for we discovered a great deal of life in the cemetery. Not only worms tunnelled assiduously through the soil, turning and aerating it, but pheasants and Canada geese raised broods among the gravestones, racoons and foxes brought up their litters in the shelter of the mausoleums designed for acres of Italian dead. How lovely, I thought, that even here in New York City you could die secure in the comforting knowledge that a racoon, warm and friendly, was going to bring up a family on your chest.

It was, I suppose, singularly appropriate that we went from this monstrous cluster of cadavers to the New York

City dump. It is a salutary experience to see what a vast quantity of waste is produced by a conglomeration of human animals all living in one spot and being as wasteful as only the human animal can be. This monstrous, simmering, multicoloured pile of garbage lay there, being added to hourly. I have been disgusted by human wastefulness frequently, for I have watched in Africa and South America people use a fragment of a tin can, a tiny length of string and a piece of paper the size of your thumbnail as a means of survival, and yet in these same countries, such as Argentina, I have looked out of my hotel window and seen the refuse-cart passing below, filled with loaves of bread scarcely touched, steaks as thick as a volume of *Encyclopaedia Britannica* with only just the centre section cut out, piles of beans and vegetables in these trundling carts that could have kept an army of Indian villages functioning for months. I have watched families in North America whom I, in my innocence, thought were suffering from some glandular disease until I discovered that this extraordinary wobbling obesity was due to overeating. What a feast they would have been if they had been Christian missionaries who had wobbled out into the outback. Of course, this gigantic garbage-heap was considered by the seagulls to be the best restaurant in New York and they turned up there in their thousands, wheeling, screaming, fighting each other, diving into the piles of refuse in search of titbits. It was, in a curious sort of way, comforting that such monstrous waste was at least going towards keeping up flying battalions of handsome birds.

So we continued to film in this, one of the most squalid, repulsive, dirty, beautiful and exciting of cities. We shot, as I have said, the wildlife in cemeteries and in city dumps, we also showed how feral dogs and cats lived in the slums, how pigeons and rats survived in the jungle of concrete, and we even showed how, fifteen or twenty floors up, in an apartment consisting of cement, glass and chromium plate, you could still find firebrats in your television set, cockroaches in your carpet and mice in your wainscoting.

Then we came to the great day known as the Battle of Block 87.

Among other people helping us, we had a charming lady naturalist called Helen Ross Russell, who had for years studied the flora and fauna of the Big Apple and had written several extremely interesting books on the subject of wildlife in a city. She knew on which skyscrapers peregrine falcons nested, where was the best place to find rats and at which golf course racoons habitually stole all the golf balls. With this fund of esoteric knowledge, her assistance was invaluable. One of the things we wanted to show was the quantity of life that could be found in what is known in America as a vacant lot and in England would probably be called a bombsite. Even in the midst of great cities, it is astonishing how nature creeps back. Moss and lichens are generally the first to appear, followed by weeds, and then even trees start to sprout between the bricks and rubbish. As the plants get a hold, the invertebrates move in – the millipedes, the spiders, the snails – and these are closely followed by various birds, mice and in some cases even toads and snakes. Thus, a vacant lot or a bombsite can, to the amateur naturalist, produce an extraordinary variety of flora and fauna.

Alastair had discovered the perfect vacant lot for our purpose. It was on the corner of 87th Street, bounded on two sides by the tall walls of buildings and on the other two sides by streets along which moved a steady stream of traffic. The lot itself was used for the most part by dog-owners exercising their pets, so it was, to say the least, well manured. Heaped with rubble, old tin cans and discarded notices – one saying 'Police Precinct' – it had provided a place for various weeds to flourish and there were even a few quite sizeable trees. There were several areas where puddles had formed and these were used by all the local pigeons and sparrows as a thirst-quenching bar-cum-swimming-pool. So our vacant lot had spiders, snails, millipedes, birds and dogs, and doubtless at night it had mice, rats and cats. It was, however, deficient in one respect – it had no tent caterpillars. This was our undoing.

Tent caterpillars are one of the major pests in America, but in spite of being such a nuisance are really quite fascinating (in the same way as human beings are). The female moth, after mating, lays an egg mass and the caterpillars form within the egg, but lie in a quiescent state until the following spring. They can endure very low temperatures by replacing some of their body fluids with a substance called glycerol, which is the tent caterpillars' equivalent of anti-freeze. When spring arrives, the tent caterpillars hatch and as a family (for that is what they are) they set about spinning a tent for themselves to live in. These tents are all-important to the caterpillars, for they act in fact like miniature greenhouses. They are oriented in such a way that they obtain maximum sunshine both in the morning and in the afternoon. Scientists have recorded that when the outside temperature was only 52° the temperature in a cluster of caterpillars residing in their silken dome was 102° Fahrenheit.

The caterpillars, as they venture from their tent to browse on the leaves of the host trees, leave a trail of silk from the spinaret under their head. As they move about the branches, they thus create little silken highways which are added to and refurbished by their brothers and sisters. However, this is only part of the story. We now come to an extraordinary piece of research, the unravelling of a sort of natural-history detective story. Scientists have stated they discovered that each caterpillar on its silken highway laid a scented trail which told its brethren which were the good routes that led to food, so in fact these silken highways were also scent guides to the best supermarket of leaves, as it were; but what intrigued the scientists was what substance it was that the caterpillars secreted in the tail end of their abdomens that acted as the guiding scent, much as a beautiful woman might leave a trail of Chanel No. 5 as she crosses a room, caterpillars' scent denoting a supply of provender, the woman's scent a possible assignation. The two things were designed along the same lines, but for different purposes. It was then that a lady researcher of great intelligence and perception called Janice Egerley made an extraordinary discovery. She observed that a caterpillar among the multitude she was studying followed a pencil line she had made in her notebook. Was there, then, something in the lead of the pencil that resembled the elusive caterpillar scent? There was, and subsequent investigations of various lead pencils proved that there was some component in the sterated beef fat or the hydrogenated fish oil that some manufacturers use to produce the lead in pencils that so excited the caterpillars with thoughts of succulent green leaves. It was further discovered that, whatever this mysterious substance was, the tent caterpillars were sensitive enough to be able to distinguish between 3B and 4B pencils of a certain brand. The investigations continue, and doubtless other fascinating facts will come to light. However, armed with this knowledge, we felt we could not depict wildlife in a city without showing some of the private life of the tent caterpillars, who are such a part of city life, being such major defoliating pests.

But, as I have said, our lovely empty lot, although it contained a young cherry tree, one of the caterpillars' favourite foods, had no caterpillars. A high-level conference divulged the fact that the budget could actually run to the extraordinary extent of getting Helen to import for us some caterpillars from her part of the city (where they were flourishing and hated) to our vacant lot, where we planned to put them on the baby cherry tree so we could film them. They would then be sternly banished back to Helen's part of the city.

So we set to work. It was here – not for the first time – that we became grateful for Paula's ability to roar, for some of the shots had to be done from the roof of a building across the busy road and so Paula's lungs and vocal cords were well exercised in shouting instructions to us and it says much for the force of her voice and clear diction that she was able to instruct us impeccably from five flights up and across an endless stream of juddering, roaring, honking traffic below her. So we finished most of the major scenes and then came to the tent caterpillars. With reverence they were removed from the van, each branch of the cherry on which they rested carefully enshrined in muslin. Carefully, we carried them over to where our cherry tree, twisted and deformed – like a child in a slum – was still making a brave show of defying New York and its attempts to exterminate it. Carefully, the branches with caterpillars on them, their tent and their silken *autostradas* were taped to the branches of our cherry tree so that the whole thing looked, if anything, slightly more natural than nature. It was at this point that we noticed a lady had joined us and was watching our activities with slightly vacant, open-mouthed interest.

'What are youse all doing?' she asked, shifting her bulk uneasily in her tight pants and denim jacket.

Alastair turned and beamed at her benignly, head on one side. Fortunately before he could bemuse still further an already puzzled mind, Paula stepped in.

'We are making a film about wildlife in a city,' she said. 'We want to show how even in the depths of a city like New York nature can still be found.'

'Is that what them bugs is for?' asked the lady.

'Yes,' said Paula kindly. 'They are called tent caterpillars.'

'They don't live here, though,' said the lady. 'You brung 'em.'

'Well, yes. You see, there weren't any here, so we had to bring them for the film,' said Paula, slightly flustered by the Neanderthal stare of the lady, who looked as though she had just returned from sweeping up singlehanded the debris of the May Day celebrations in Red Square.

'If there were none here, why did you brung 'em?' asked the lady.

'For the film,' snapped Alastair, who was trying to concentrate on whether he wanted the caterpillars to walk from right to left or from left to right and whether they would obey him.

'But that's faking,' said the lady, arousing herself out of lethargy into a sort of Middle European position of argumentation, feet slightly apart, hands on hips. 'You brung 'em here, and they don't live here. That's faking. You brung them bugs here deliberate.'

'Of course we brung them here,' said Alastair irritatedly, his train of thought interrupted. 'If we had not brung them, there wouldn't be any for us to film.'

'That's faking,' said the lady. 'That's not true.'

'Do you realize, madam,' I said, in a role of peacemaker, 'that ninety per cent of the films you see on wildlife, like Walt Disney, are faked? The whole process of filming is in a sense a fake. However, no more than a portrait painter or a landscape artist fakes, that is to say he rearranges nature to a better angle for his purposes.'

'Walt Disney doesn't fake,' said the lady, now starting to show all the belligerence of a sabre-toothed tiger immersed in a sort of intellectual tarpit. 'Walt Disney is an American. What youse is doing is faking, and faking on our lot.'

'We have permission from the Mayor's office,' said Paula.

'Have you got permission to fake from the 87th Street Block Association?' asked the lady, swelling as a turkey to a gobble.

'Surely the Mayor's office takes precedence?' asked Paula.

'Nothing takes precedence over the 87th Street Block Association,' said the lady.

'You know . . . for some . . . skyscrapers . . . lots of life . . . caterpillars . . .' said Alastair, turning in a distraught circle.

'I'll go and see the 87th Street Block Association,' said the lady, 'and find out why you are allowed to fake.'

She strode away, as if to relieve Leningrad singlehanded, and we all heaved a sigh of relief. However, our sense of relief was short-lived. Alastair was shouting instructions at a tent caterpillar who was incapable of understanding English when the lady returned bearing with her a woman who looked like one of those viragos who have been hatched from a shrike's egg, with militant eyes like laser beams, the sort of person who always looks for the worst in anything. Accompanying her, presumably as a back-up, was a man who appeared to have been constructed out of cardboard at a very early age, and rained on incessantly during his life.

'What is going on here?' asked Shrike Lady.

Patiently, Paula explained to her about the film we were attempting to make, while Alastair continued to turn in irritated circles.

'But what are you doing to our lot?' said the lady accusingly, rather as if the place was Kew Gardens instead of a vacant lot knee deep in dog droppings.

'They're faking nature,' said Neanderthal Lady. 'They've brung a lot of bugs.'

'Bugs?' said Shrike Lady, her eyes flashing. 'What bugs?'

'These,' said Alastair, pointing. 'They're only tent caterpillars.'

'Tent caterpillars?' screamed Shrike Lady. 'You brought tent caterpillars to our lot?'

'Well, there weren't any here,' said Paula.

'Yes, and we don't want them here,' said Shrike Lady.

'But we only brought them for the sake of the film,' said Paula. 'We'll take them away again.'

'We don't want our vacant lot covered with tent caterpillars,' said Shrike Lady, her voice taking on a rasping quality.

'Disgusting,' said Cardboard Man. 'I have been a journalist for twenty-five years, and I have never heard of such a thing as faking nature.'

'If you've been a journalist for twenty-five years, you must have come into contact with a considerable quantity of untruths,' I said with some asperity. 'Surely you must realize that practically every nature film that you have ever seen has been faked in one way or another.'

'He said Walt Disney was a fake,' said Neanderthal Lady, apparently as heinous a crime as lighting a fire with a piece of the true Cross.

'Disgusting,' said Cardboard Man. 'No real journalist would stoop to faking.'

'*And* accuse Walt Disney,' said Neanderthal Lady.

'Dear God,' groaned Alastair, 'we are losing light.'

'All we have done,' Paula explained patiently, 'is to tie two branches on to this cherry tree on which there are some tent caterpillars. When we have finished taking the pictures of them—'

'When you have finished *faking* pictures of them,' said Cardboard Man indignantly, 'a thing which no true journalist would do.'

'When we have finished taking pictures of them,' repeated Paula, 'we will remove them.'

'Where will you take them to?' asked Shrike Lady.

'Back to where they live,' snapped Alastair, 'and a damn sight more salubrious part of New York than this.'

'What is wrong with 87th Street?' asked Shrike Lady.

'Yes, and who are you to come here criticizing the 87th Precint?' said Cardboard Man. 'And you an Englishman, too, or from Boston maybe.'

'Look,' said Paula, 'we will shoot these shots in about five minutes, and then we are going to pack up everything and leave your vacant lot vacant.'

'We can't have our vacant lot used in fakery,' said Shrike Lady. 'It's our vacant lot.'

'But we aren't doing any harm,' said Paula. 'After all, the people who bring their dogs in here do more harm, I reckon.'

'You brung bugs in,' said Neanderthal Lady, 'and before we know it there will be bugs all over 87th Street.'

'Dear God,' groaned Alastair, 'this is ridiculous.'

'It may be ridiculous to you, but it's not ridiculous to us,' said Shrike Lady. 'You come here and pour tent caterpillars all over our vacant lot and expect us to take no action.'

'Do you think that Alfred Hitchcock had problems like this?' I enquired of Alastair.

'I demand that you take your bugs away,' said Shrike Lady.

'I agree,' said Cardboard Man.

'We are going to take them away,' shouted Alastair, 'as soon as we have filmed them.'

'We are not having any faking on our vacant lot,' said Shrike Lady.

This ridiculous conversation went on and on until the three characters made our lives so unbearable and the loss of light so dangerous we were forced to pack up our tent caterpillars in the van. Only when she had watched us put them into the van and they were safely under lock and key was Shrike Lady satisfied.

Frustrating and annoying though this was, looking back on it one felt the incident had a certain charm. It was nice to feel that in that giant, brash, apparently uncaring city there were people willing to take up cudgels about a vacant lot covered with dog droppings.

SHOOT TEN

FROM THE BUSTLE and hurry of New York it was pleasant to return to Europe and, moreover, work in one of my favourite countries, Greece, with its blue sea and skies, its sparkling clear air that makes it one of the most magical of places.

Typical of the topsy-turvy way television filming is conducted, the programme that was to be the first in the series was filmed last. As the whole theme of the series was about how to become an amateur naturalist, what better place to start (thought Jonathan) than the island of Corfu where I had first developed my passion for animals when I lived there as a child. I was pleased with this idea since I had not visited the island for many years in spite of all my friends there imploring me to do so, and Lee had never been there.

The island, in spite of the tourist trade and all its attendant vulgarities and desecrations, still retains a special magic, and I was anxious to show Lee those bits of it that still remain the same as they had been when I was a boy. We were doubly lucky inasmuch as my friend Ann Peters was living on the island, spoke Greek fluently and offered to put herself at our disposal. At one time, Ann had been my secretary and had accompanied me on filming trips to Sierra Leone and complicated conservation trips to Australia and Patagonia, so she knew not only the difficulties of filming but also the particularly horrendous problems of filming animals.

'Where are we staying?' I asked Jonathan.

'The Corfu Palace,' he answered.

I stared at him in disbelief. The Corfu Palace was the oldest and most venerable of the island's hotels, built in the early days of the century. It had been cunningly arranged (as only an architect worth his salt could arrange it) on a sweeping bay on the outskirts of the town, where the entire sewage system of Corfu was debouched into the sea. This gave the whole area – especially in summer – such a rich aroma that even dogs avoided it.

'Who suggested this?' I asked.

'Ann,' said Jonathan.

I stared at Ann, thinking that a prolonged sojourn in Corfu had addled what brains she used to possess.

'Have you taken leave of your senses?' I asked. 'First of all, we will all be asphyxiated in our beds; second, the rooms will cost you ten billion drachmas a second; and, third, if we want to keep butterflies or tortoises handy, you can't do it in a hotel of that aristocratic decrepitude.'

'It has all been arranged,' said Ann soothingly. 'First, the manager – a very nice man called Jean-Pierre – is giving us special rates; second, the sewage problem has now been solved; and, third – and this is really something – Jean-Pierre is a mad-keen herpetologist.'

I took a deep, steadying draught of B & S before replying.

'Now I know you have gone round the twist,' I said with conviction. 'I know Corfu is eccentric, but I refuse to believe that even there you can find a herpetologist in charge of one of the best hotels in the island.'

'But it's true,' Ann protested. 'He has got a flat on the top floor of the hotel and he keeps snakes and tortoises and all sorts of lizards up there and moreover he has offered to go out and catch any reptiles we want to film.'

I gave up. The island of Corfu was in the past as packed tight with eccentricities and surprises as a magician's trunk and I could see it had not lost its power to surprise.

The island lies like a strange, misshapen dagger in the blue Ionian Sea, midway along the Greek and Albanian coastlines. In the past, it has fallen into the hands of a dozen different nations, from all of which it has absorbed what it found good and rejected the rest, thus keeping its individuality. Unlike so many parts of Greece, it is green and lush, for when it was part of the Venetian empire they used it as their oil store, planting thousands of olive trees, so that now the bulk of the island is shaded by these carunculated giants with their wigs of silvery-green leaves. Between them run the admonishing fingers of black-green cypress, many planted in groves as dowries. All this creates a mystical landscape, bathed in sharp brittle sunlight, orchestrated by the knife-grinder song of the cicadas, framed in the blue, still sea. Of all the wonderful and fascinating parts of the planet I have been privileged to visit, Corfu is the nearest approach to home for me, since it was here, nurtured in sunlight, that my fascination for the living world around me came to fruition.

Owing to the vagaries of connecting flights, we managed to stay for a few hours in Athens – enough time for us to catch a brief glimpse of the Acropolis, a quick peep at the Evzones changing the guard at the royal palace and then some time eating a splendid meal on the waterfront in

Pyraeus: seafood as only the Greeks can cook it. Then we flew on to Corfu.

When we reached the island, it was dark and a giant yellow moon lit the road so brightly you could see the olive-enshrouded landscape clearly and the moon's reflection on the faint, wind-stirred sea was like a million buttercup petals on the surface. After an excellent bottle of pale amber retsina, tasting of all the pinewoods you had ever visited, and some delicious local fish, we retired to bed and not even the moon perched, it seemed, on our balcony rail could keep us from sleep.

The next morning we were having breakfast when Jean-Pierre arrived and was introduced. Short and dark, he had humorous brown eyes and a delightful smile. To the alarm and consternation of the other guests breakfasting among the flowerbeds, he produced out of several cloth bags one of the largest grass snakes I had ever seen in my life, a beautiful glass snake that looked as though it were cast in bronze and then, with a final gesture like a conjuror producing a rabbit out of a hat, he poured out from a bag on to the flagstones of the patio a cascade of European pond terrapins, dark greeny-black, yellow-spotted, with golden eyes like leopards.

'This is all I have caught for you, I'm afraid,' he said apologetically, and the pond terrapins came to life and scuttled off among the tables. After a breathless five minutes, they were rounded up and put back into their bag.

'Where did you catch them?' I asked.

'I got up very early and went to a lake called Scottini,' he replied. 'It is in the centre of the island.'

'Oh, I know it well,' I said. 'It was one of my favourite collecting areas.'

'It is an excellent place for all sorts of wildlife,' Jean-Pierre said.

'We are going to film a sequence there with pond terrapins,' said Jonathan, who had, some weeks previously, done a quite extensive reconnaissance trip to Corfu. 'We will see you approaching the lake and there will be all these terrapins on the bank. Slowly, you and Lee will approach them. They will indulge in terrapin behaviour like—'

'Like sitting stationary on a rock,' I said. 'Have you told them? Have you given them each a script? Have they read their contracts? I refuse to act with a bunch of pond terrapins that haven't read their contracts and won't do what they are told, won't take direction and – worst of all – who forget their lines. Remember, my reputation is at stake, too.'

'Well, anyway,' said Jonathan, giving me a Heathcliffian look, 'I am sure they will be perfect.'

'Where are we going to keep them?' I asked.

'Why not in your bath?' suggested Jean-Pierre seriously.

It was as though the manager of Claridges or the Waldorf Astoria had suggested you kept a group of warthogs in your suite.

'That's a good idea,' said Lee, 'and we can take them out when we have a shower.'

'Yes,' said Jean-Pierre, 'they don't like soap and hot water.'

It was, I decided, the sort of conversation you could only have in Corfu. So we went up to our room, carrying our reptile stars, and filled the bath with water and put the terrapins in. The snakes we knew were all right in their bags. Then we set off for the very north of the island to a place called Kouloura where Jonathan wanted to film our 'arrival' in Corfu on board a caique, one of those tubby, highly coloured Greek fishing boats that are such a feature of the Greek landscape.

It was a brilliant blue day, as clear as crystal, with a sun that was just comfortably hot. The sea was blue and calm, and only the faintest whisper of wind came from the brown eroded hills of Albania and the Greek mainland that we could see so clearly across the waters. It was cool driving through the olive groves in the shade created by the great canopy of silvery-green leaves and the huge twisted olive trunks, pitted like pumice-stone, each unique as a fingerprint, looking like misshapen columns holding up the roof of a cathedral of leaves. Soon we left the coolness of the olive groves and started along a road that twisted and turned along the flanks of the biggest mountain of Corfu, Pantokrator. Here in places the edge of the road dropped almost sheer to the glittering sea, and above, the rocky slopes of the mountain rose; among its russet, gold and white cliffs the red-rumped swallows glided like dark arrows, busy about the construction of their strange nests that look like half Chianti bottles made out of mud cobbles.

Presently, we took a steep, winding side-road towards the sea, a road thickly lined with immensely tall dark-green cypresses that had been elderly giants when I used to come here in 1935. Soon we could see below us Kouloura harbour, like a small curved bow, and at one end of it what is probably the most beautiful villa in Corfu, belonging to old friends of mine, Pam and Disney Vaughan-Hughes. Anchored in the harbour was our caique, a splendidly large craft, spotlessly clean, its blue and white paint gleaming.

Pam and Disney greeted us warmly, for we had not seen each other for several years. They had kindly agreed to have our gear stacked in front of their beautiful home, to let us film in their lovely garden, to ply us with cool drinks and even to lend us the talents of their land tortoise, who rejoiced in the name of Carruthers. Friendship could go no further. So we cluttered up the front of their house as only a film crew can and while the team was setting up we went to have a look at the kayiki to make sure all was ready for our sea trip. Here, to Jonathan's horror, disaster struck.

In the opening sequence on the boat, I was to say: 'All of us are born with an interest in the world around us. You watch any young human being – or any other young animal, if it comes to that – and you'll see that they're investigating and learning the whole time with all their senses. Because from the moment we're born we are explorers in a very complex and fascinating world. Now, as people grow older, they sometimes lose interest in the world around them, but others keep it stimulated the whole time. These are the lucky ones. These are called the amateur naturalists.'

In order to make the point more forcefully, Jonathan had decided we needed a child on board with us, so that we could all be examining a big bowl full of sea-creatures. To this end, he had engaged the services of the daughter of the owner of the minute café that graces Kouloura harbour, a very pretty little six-year-old. However, just before our arrival she had done something so monstrously naughty (we never found out what) that her mother had taken the unprecedented step (in Greece, that is) of giving her a good slapping. The result may be imagined. Jonathan found his tiny talent in tears, sitting in a mournful heap, refusing to speak, refusing to get into her best dress, refusing everything. In vain did Pam, Ann and I – the only three Greek-speakers among us – cajole and flatter and beseech. Even Jonathan's munificent offer (with total disregard for the budget) of raising her fee from ten drachmas to twenty had no effect.

'We can't do the scene without a child,' said Jonathan. 'For heaven's sake, *do* something, Ann.'

'What do you expect me to do?' asked Ann. 'If the child won't do it, you can't make her.'

'Then, find someone who will,' snapped Jonathan.

So poor Ann was dispatched to the nearest village as a talent scout.

'Does it have to be a she, or will a he do?' she asked before she left.

'I don't care if it's a hermaphrodite as long as it's a child,' said Jonathan, glowering.

For the next half-hour, while we waited for Ann to return, Lee and I investigated the warm shallow water in search of props in the shape of choleric-looking hermit crabs inhabiting brightly coloured topshells, other shells containing their rightful owners and large spiky spider crabs, each wearing its coat of weed and sponges on its back which these creatures plant on themselves to escape detection. The collection of this vast array of living props slightly mollified our director, though he was still twitchy and we awaited Ann's return anxiously.

Presently, she came back triumphantly bearing with her a good-looking little boy of about ten. As the car with this male talent drew to a halt, out of the café door stepped the little girl wreathed in smiles and wearing her new dress.

'Gee, honey, look,' said Paula excitedly. 'Now you have two children.'

'Do you think the budget will run to two?' I asked Jonathan seriously. He just glowered at me.

So we spent the rest of the day filming the caique sequences, which were rather complex, for as well as the shots on board (which were difficult but not too bad) Jonathan wanted to go to a high vantage-point on the mountainside and film a panoramic view of the harbour and Pam and Disney's house with the caique chugging majestically into the harbour. As we had no walkie-talkies, we had to accomplish these shots by Jonathan reaching his vantage-point on the mountain and I watching him carefully through my binoculars while the kayiki went round and round in tight circles awaiting instructions. When Jonathan waved his arms we straightened out and went into harbour. Needless to say, because this was a most complex shot we had to do it several times. Eventually, Jonathan was as satisfied as a director ever is and we got packed up and set off on the long, hot drive back to town, our minds full of thoughts of icy drinks, clean clothes and delicious food.

The terrapins were still in the bath.

The next day, disaster of another sort struck. Jonathan had tracked down one of the villas that I and my family had lived in when we were in Corfu and had decided that it was most photogenic and suitable for a long sequence in the film. After numerous and complicated telephone calls, Ann had managed to track down the owner in Athens, in order to obtain his permission to film in and around the villa, only to find that the villa was rented to a nightclub owner, whose permission would also be necessary. It was even more difficult to track down the nightclub owner than to track down the real owner of the house since nightclub owners appear to be crepuscular, or at any rate nocturnal, and so during the day they are unavailable and the moment it gets dark (like Dracula) they leave their coffins and flap to and fro around the city, making it exceedingly hard to make contact with them. Finally, Ann managed to find him in some undiscovered tomb, whereupon he said that in no circumstances would he allow us to film in the villa. After prolonged pleading, Ann finally got him to agree to allow us to open up the house, but only if he were there. He told her his date of arrival in Corfu and said that he himself would open up the villa for us. Alas for Jonathan's nerves; the day came and went and the man was conspicuous by his absence.

'Why don't we go up there and you can at least film in the grounds and on the veranda,' Ann suggested sensibly. 'Maybe he will arrive tomorrow.'

'Well, I suppose so,' said Jonathan moodily, 'or we could film at Potámos in the hope that he will be on tomorrow's plane.'

So we went to Potamos, a charming village, straggling up a hillside, the neat, multicoloured houses with their arched verandas exactly as I remembered them from forty years ago. Under every arch was a swallow's nest full of gaping young and under every nest was a cardboard box to catch the faeces so amply and generously shared with you by the swallows. I was reminded of the Greek saying that a house is not a home until it has a swallow's nest under its eaves. As I watched the parent birds, beaks stuffed to overflowing with insect provender, skimming in to hang on the nest and stuff the gaping mouths of their young, I thought that these were probably the great, great, great, great, great, great, great, great, great, great-grandchildren of the swallows I watched under these identical eaves when I was a child. After we had filmed them and some other sequences in the village, we returned to the hotel.

The terrapins were still in the bath.

The next day dawned bright and clear. The plane from Athens arrived and our man was not on it.

'To Hell with him,' snarled Jonathan. 'We'll go up to the villa and film anyway.'

The villa was one I described in the book I wrote about my childhood in Corfu which I had called the Snow White Villa. It lay in a large and ancient olive grove and was shaded by a huge magnolia tree, oleanders with pink and white flowers and a grape vine over the veranda that in season was heavy with bunches of white, banana-shaped grapes. Alas, when we drove up the rock-strewn pot-holed drive and drew to a halt outside, I could see that the villa was snow white no longer. Its once white walls were discoloured with patches of damp, there were huge cracks in the plaster, and the green shutters were faded with the paint peeling off them. In spite of this, the villa still somehow managed to look elegant, even in decay, but I wondered how anyone could treat such a beautiful and charming building in such a brutal way.

As the equipment was unpacked, I led Lee around the overgrown garden among the olive trees and indulged in nostalgia. Here was the veranda where, at one of our numerous parties, my various animals had caused havoc; my magpies escaping, getting drunk on spilt wine and then wrecking the carefully arranged table just before the guests arrived, while beneath the table lurked my fearsome gull, Alecko, who bit the guests' legs as they sat down to eat. This was the wall in which my favourite gecko, Geronimo, used to live who fought to the death on my bedroom wall the praying mantis twice his size. About a hundred yards from the villa there stood the family chapel, one of those charming miniature churches that you so frequently find dotted about the Greek countryside, built God knows when and dedicated to some obscure saint who had performed some miracle or other. This one was painted pink outside and was the size of a large room with curious fold-back seats for the congregation and at one end, over the altar, a picture of the Virgin Mary and Child. Now it was all faded and forlorn, and a drift of old winter leaves had half wedged open the doors and spread in piles across the floor. In my day, the floor had been swept and garnished, the seats polished, and two tiny oil-lamps had been constantly burning with just enough light to illuminate the Virgin and Child, and fresh flowers were always kept in a vase below the portrait. Now all smelt of decay and there were no lamps to light and no flowers. I remember once returning from some expedition of mine after dark and I saw that the doors of the little church had been accidentally left open. Putting down my butterfly-net and collecting-bag, I went to close them and came upon an astonishing sight. It was the season of the fireflies, and when I got to the doors and looked into the little church there was the picture of the Virgin illuminated – seeming almost to float – in the daffodil-yellow light of the tiny oil-lamps, but as well there were dozens and dozens of fireflies that had flown in through the open doors and now drifted like greeny-white flashing stars around the interior of the church and others crawled over the seats and walls. A few had landed on the Virgin's portrait and decorated it like some pulsating, moving jewels. Enraptured, I watched this beautiful and eerie sight for a long time and then, fearing to lock the fireflies in the church in case they died, I spent an exhausting half-hour catching them with my butterfly-net and releasing them, and as I did so I felt that the portrait of the Virgin must have been sorry to lose such an exquisite decoration to her church.

When we returned from the church we found Jonathan wearing a smugly guilty expression, holding a piece of glass in his hands.

'Look,' he said, holding it up. 'I was just looking through one of the back windows when this piece of glass fell out. If I put my hand in, we can open the window and we can get into the villa.'

I sighed. 'I'm not sure what the sentence in Greece is for breaking and entering, but I think it would be several years and Greek gaols are notoriously uncomfortable.'

'Surely nobody would know,' said Lee, 'not if we put the glass back.'

'And what happens if the owner turns up while we are disporting ourselves in the villa?' I asked.

'Let's cross that bridge when we get to it,' said Jonathan firmly. 'At least we can get the veranda scenes done.'

So we broke into the villa. The reason we needed to get into it and open it up was twofold. First, we needed shots of me and Lee looking inanely out of various windows and going in and out of the door and, second, we needed the electrical supply within so that we could film the night-time and evening sequences. All of this took some considerable time and it was quite late in the evening when we finally packed up and (having carefully locked up the villa and replaced the glass) wended our weary way back to the Corfu Palace.

The terrapins were still in the bath.

The next day we had a free morning as Paula and Jonathan had to go and clear caterpillars through Customs. I feel somehow that this statement needs a certain clarification. Owing to the fact that we were filming in Corfu at the wrong time of the year for butterfly larvae (at least for the ones we wanted to show), we had been forced to import the caterpillars from a well-known butterfly-farm in England. Needless to say, the Greek Customs thought that, although Corfu is renowned for its eccentrics, it was carrying things too far when they insisted on opening this highly suspicious parcel to find it full of fritillary, cabbage white and swallowtail larvae all nestling in beds of their favourite food plants. Arguments by Ann that all these species were found in Corfu anyway met with cold unfriendly stares from the Customs officers. Why, they asked, if they were found in Corfu already, did one have to import them at colossal expense from England? (The complications of animal filming are difficult, if not impossible, to explain to a Greek Customs officer.) In any case, they pointed out – national pride now coming into it – if the same caterpillars were found in Corfu, why not use them? Were Greek caterpillars inferior in some way to British caterpillars? For years, Greece had been known for the enormous size, quantities and qualities of its caterpillars. Greek caterpillars, as everyone admitted, were the best in the world. Therefore, what was the sense in importing a lot of English caterpillars (inferior in every way) and probably causing some fatal disease to the Corfiot caterpillars? The morning dragged by, while Paula, Ann and Jonathan, hot and irritated, had to sign affidavits to the effect that the caterpillars had been individually inspected by the Royal College of Veterinary Surgeons, the Royal College of Surgeons, the Ministry of Agriculture, and the Zoological Society of London. Further guarantees were signed to indemnify the Greek government, and a promise to pay vast compensation should the English caterpillars be responsible for one single (superior) Corfiot caterpillar's death. Further guarantees had to be given that our caterpillars were not in any way to corrupt the Corfiot caterpillars by coming into contact with them and that our caterpillars, at the end of the filming, would be ignominiously banished from Corfu and sent back to England – there, presumably, to hatch out into inferior butterflies. The three of them arrived back at the hotel limp with exhaustion but triumphantly bearing our English caterpillars in their midst.

So that afternoon we went back to the villa. The pane of glass once more mysteriously fell out, the villa was opened up and we commenced work. Caterpillars as co-stars leave a lot to be desired, like most animals. Either they remain as unmoving as museum specimens, or else they gallop about their respective food plants at such speed that it is difficult to follow them with the camera. After we had finished with the caterpillars we had to try to film all the other insects that Lee, Ann and I had been assiduously collecting over the last few days – tiger beetles, scarabs, cicadas and the like. These were all carefully incarcerated in a series of pots and matchboxes. When I was a boy in Corfu, I did not have access to all the refined collecting apparatus which is accessible to the young naturalist today, so necessity had to be the mother of invention in no small measure. In our villa, not a pot or jam-jar was thrown away; cardboard boxes were gold dust, as were tins, but best of all were matchboxes, for they were light and easy to carry and were small enough to prevent your capture from rushing about and hurting itself. I must have had a collection of several hundred at one time which I would take out with me on my expeditions and return with them full – a sort of matchbox menagerie. Of course, matchboxes, though extremely useful to me, sometimes led to trouble. I vividly remember the day I inadvertently left one of my matchboxes on the mantelpiece and my elder brother (no animal lover at the best of times) opened it to get a light for his cigarette, whereupon a female scorpion covered with young crawled out on to his hand. I will draw a veil over the ensuing proceedings. However, for the making of this film, I was delighted to find that the humble matchbox had not lost its usefulness; it was merely the occupants that proved recalcitrant, either falling off the flower on which they were put or promptly unfurling their wings and flying away, so by the time we had got to the point which is called 'loss of light' (that is, it got so dark that even the director admitted, reluctantly, that one could no longer film) we packed up our inferior caterpillars and returned to town.

The terrapins were still in the bath.

The next day dawned (as all the days had done) sunny, with a halcyon blue sky. As we sat lingering over our breakfast, Jonathan appeared, apparently in an unprecedentedly good mood. He ordered his usual meagre breakfast of cereal, coffee, toast, marmalade, sausage, bacon, fried eggs and fried potatoes with fruit salad and cream to follow and then sat back and beamed at us.

'Today,' he said, as one offering an elephant ride or similar treat to a child, 'we are going to film the snake sequence. Jean-Pierre is coming to help.'

'You mean the glass snake?' I asked.

'Yes,' he said, 'the glass snake.'

'May I point out that "glass snake" is a misnomer?' I explained. 'A glass snake is in fact a large, legless lizard.'

'But it looks like a snake,' protested Jonathan, not for the first time annoyed at Mother Nature's perfidy.

'Nevertheless,' I said, 'it has vestigial limbs and if you handle it roughly its tail will come off like a lizard's. Hence the term "glass snake".'

'Oh God,' said Jonathan, aghast, 'that's all we need, for the bloody thing's tail to drop off in the middle of a sequence.

Why the Hell am I making animal films?'

So with Jean-Pierre and our reptilian star we set off to an olive grove that Jonathan had decided was the most photogenic one on the island, though how he could choose since all the olive groves were magnificent it was difficult to understand. I had informed him that the glass snake or sheltopusik (to give it its European name which slips so easily off the tongue) moved with the speed of light and so should be filmed in an area where recapture was made easy. Jonathan assured me that he had taken that into consideration. To my surprise he had, for when we got to the olive grove we saw running along one side of it a donkey track guarded on each side by a rough dry-stone wall forming, as it were, a meandering trough suitable for the release of snakes, though not having been designed for that purpose.

'Now,' said Jonathan, 'I want you and Lee to start way down there by that olive and when you reach that bush there you suddenly see the glass snake and catch it.'

'Wait a minute,' I protested. 'If we are supposed to walk fifty yards before we get to the creature, it will be about five miles away.'

'Well, what do you suggest?' he asked.

'It must be released just as we reach the bush,' I said.

'How?' asked Jonathan.

I gazed down the path at the place where the capture was supposed to take place. At that particular spot, the meandering of the path formed a slight bend and in consequence the wall took a bend that formed a tiny enclave.

'If somebody is concealed there at that point, he can release the creature as we get there,' I said.

'Who?' asked Jonathan.

'I will do it,' said Jean-Pierre, herpetologist manager of the Corfu Palace Hotel, now stripped to the waist and ready for action.

'OK,' said Jonathan. 'Go down there and let's see how it looks.'

Obligingly, Jean-Pierre, carrying the bag with the glass snake in it, trotted down to the bend in the path and stood there.

'That's no good. We can see you,' shouted Jonathan. 'Get down a bit.'

Obligingly, Jean-Pierre got down on his haunches.

'It's still no good,' shouted Jonathan. 'We can see your head. Lie down.'

So the manager of the Corfu Palace stretched himself face downwards in the dust behind the wall. If his clients could have seen him at that moment, it would have given them pause for thought.

'Excellent,' shouted Jonathan. 'Just stay there and release the glass snake just before Gerry and Lee get to you.'

So Jean-Pierre lay face downwards in the dust, the sun beating down on him while we had a couple of rehearsals. Then all was ready. At the crucial moment, the glass snake was released and, to my astonishment, behaved in an exemplary manner, speeding across our path and then curling itself up in a clump of grass under the wall where it could be easily captured without risk to its tail. Jean-Pierre rose to his feet, his back covered with sweat and his front covered with fine white dust and a proud grin on his face. After we had taken all the close-ups of the handsome glass snake, his burnished body, his scales like little bronze bricks, his vestigial hind limbs, his handsome head with fine eyes and a mouth set in a broadly benign smile, we set him free and watched him slide off through the bushes as smoothly as oil. By this time we were losing light, so after a pause to eat a refreshing watermelon, pink as a sunset cloud, we packed up and went back to the Corfu Palace, taking with us its manager, hot, tired, dusty but triumphant at this, his first cinematic experience.

The terrapins were still in the bath.

The next day at breakfast Jonathan was again in a jovial mood as he ploughed his way through his provender.

'Today,' he announced, through a mouthful of sausage and egg, 'today we are going to film at the lake – you know, Scottini.'

'What are we going to film there?' I asked.

'Pond terrapins,' he said with relish.

'Pond terrapins?' I asked incredulously.

'Yes,' said Jonathan, and then, misinterpreting my expression, in alarm: 'You've still got them, haven't you?'

'Oh yes,' I said. 'I was just thinking that our bath won't be quite the same without them. We have grown quite attached to them.'

'Good,' said Jonathan. 'I thought they might have escaped or something.'

'Unfortunately no,' I said.

'Well,' said Jonathan, 'this is what we will do. We will get a shot of you and Lee arriving at the lake and you can tell how you used to go there and catch things when you were a child. We will have shots of the frogs and newts and that grass snake that Jean-Pierre caught and shots of the pond terrapins. You will explain to Lee how easy it is to catch pond terrapins—'

'I beg your pardon,' I interrupted. 'Have you ever seen a pond terrapin move?'

'Oh, that's all right,' said Jonathan, airily dismissing this quibble. 'After all, we've got eight. One of them is sure to be an octogenarian and have to move more slowly than the rest.'

It was Jonathan's pathetic faith that sooner or later he would find an animal that would take direction. Throughout the series, he clung doggedly to this belief, though it was never justified by circumstances.

'None of the ones we've got is an octogenarian,' Lee said

to me *sotto voce*. 'When we take them out of the bath to have a shower, they dash about the bathroom like mad things.'

'I know,' I said. 'Don't let's undermine Jonathan's pathetic faith in Mother Nature. Who knows, we might have a miracle.'

So began the great pond terrapin day. The terrapins were removed from our bath (doubtless to the relief of the maids who serviced our room) and placed in a suitable container, and our convoy of three vehicles set off, for Jean-Pierre flushed with his cinematic success as a glass-snake wrangler had come along to wrangle the grass snake for us. It was an excessively hot day and we were glad when at last the road plunged into the shady shimmering depths of the vast olive groves. These groves to me, when I was young, were magical places. To grown-ups who walked there among the trunks with their gaping holes and the canopy of silvery-green leaves, they were merely scenically beautiful and they were grateful for the shade they provided, but to me they were a treasure trove of creatures. The myriad holes in each tree provided sanctuary to a dozen different creatures from Scops owls to squirrel dormice, from wrens to black rats. At the right time of the year, you could find crawling up their trunks strange hump-backed, bulbous-eyed creatures, newly emerged from the earth. Watch them, and their skin would split down the back and slowly, and with great effort, there would emerge a cicada, with nut-brown body and silver wings, the true harbingers of summer who would make the island vibrate with their song. In the roots of the olives, you could find centipedes as long as a pencil or toads with silvery skins blotched with green so they looked like those medieval maps of the world where the continents were all misshapen. Insects were everywhere, butterflies, ant lions and ladybirds, fragile lace-wing flies who laid their eggs on slender stalks on the plant stems, and jet-black scarab beetles in pairs, rolling their balls of dung to bury as nurseries for their young. Someone once said to me that they could not understand what I saw in the olive groves – they were so dull and lifeless. For me, they housed an endless, fascinating pageant of creatures and in spring they were awash with flowers, as if someone had emptied a paintbox among the great, dark, gnarled trunks. They were anything but dull and lifeless.

Finally, we bumped our way down a stony track through the olives and there lay Scottini, an almost circular lake about seven or eight acres in extent, surrounded by trees and with a large reedy island in the middle and its shallow water full of jade-green weeds. As olive groves looked to some people, it too looked lifeless, yet I knew it was a universe of its own, for in its depths were weird darting, rolling, jerking, flitting forms of microscopic life, fearsome dragonfly larvae, small fish, newts, frogs, snakes and pond terrapins. I remember in my youth I had gone here once and spent a day collecting, and so rich was this little lake and so numerous my captures that I had soon used up all my collecting gear and was forced to use my clothes in which to carry my precious specimens, so that I arrived back at our villa stark naked, to the alarm and consternation of my mother.

So, after Jonathan had prowled around and found suitable locations, cameras were set up and we took the first sequence which starred the grass snake. Our snake wrangler, the cares of catering and hotel management forgotten, bare-footed, trousers rolled up, stripped to the waist, danced about in mud and water, getting the handsome reptile to do Jonathan's bidding. The snake behaved beautifully, slithering across mud, wriggling through grass and finally swimming out into the lake, its large handsome head raised high above the water, leaving a wide V of ripples behind it.

'Now for the *pièce de résistance*,' said Jonathan, overexcited by our success with the grass snake. 'The pond terrapins.

Now, I want you to put three of them just there on that grass bank, and you and Lee come along and see them basking in the sun and you creep up and catch one as the others go into the water.'

'They will be down that bank before you can say "Jonathan Harris",' I said.

'Well, anyway, let's try it,' said Jonathan stubbornly.

So three of the pond terrapins were taken out, held in position by Jean-Pierre, while Lee and I took up our positions.

'Now, action,' said Jonathan.

Jean-Pierre released his hold on the terrapins and leapt back so that he was out of shot. Lee and I took one step forward. The three terrapins, like racing cars taking off at Le Mans, sped down the bank, plopped into the water and disappeared.

'Damn,' said Jonathan. 'We'll have to hold them further back.'

'Remember you have only got five left,' said Lee.

'Well, we will just try another three,' said Jonathan. 'I'm sure we will get it this time.'

So the three terrapins were placed well back from the lake's edge, Jean-Pierre holding them firmly until Jonathan shouted 'Action'.

This time, the terrapins behaved differently. Apparently, they could not see the water and so they did not know which way to run. They revolved round and round for a second and then ran straight for the camera and through the tripod legs. Time and again, we tried to get them to run towards the lake, and time and again they ran inland with an obstinacy that can only be displayed by a terrapin or a donkey. Desperately, we moved them to a vantage-point where they could just see a glint of water, whereupon they sped into it and disappeared with the same alacrity as the first three had shown.

Jonathan was now wearing his most Heathcliffian scowl.

We tried again with one of the two remaining terrapins and he put up a new variation of the act. He pulled himself up into his shell and remained there immobile as a stone. Nothing we could do would make him move. Then, while we were all having a conference about his stubbornness, he quite unexpectedly came to life and rushed down the bank to freedom before any of us could stop him. Now we had only one terrapin left and things were getting desperate. Jonathan was taking no chances, so we filmed the capture scene in reverse, that is to say the first shot was Lee with the net with the terrapin in it, pulling it out of the weed as if she had just caught it. Then we released the terrapin in shallow water and filmed him as he swam away and then filmed Lee and myself rushing down the bank and performing an imaginary capture. Surprisingly, when all these bits of film were carefully cut and edited and arranged in juxtaposition with each other, it was surprisingly effective but it is a bit nerve-racking doing a scene in this way, as you are never sure until you see the shots in the cutting-room whether it is going to be successful or not. So on this last day of filming we had released all our animals and we packed up and left the little lake, placid among the olive trees, and returned to town, tired but cheerful.

There were no terrapins in the bath.

We had luxurious baths, and Lee and I spent the evening wandering about the old, narrow streets of Corfu, visiting various friends of mine, drinking far too much retsina, singing songs, eating roast lamb and fried shrimps. We knew that ahead of us lay weeks in Toronto, incarcerated in airless cutting-rooms and sound-booths, watching yards and yards of film over and over again, writing the commentaries and then reading them to film – tedious, boring work that has to be indulged in before the film is a finished product. But that lay ahead, and so this night we enjoyed ourselves.

It was very late when we walked back along the seafront to the hotel. The moon was high and clear, her craters and mountains showing as faint shadows on her face, making her look somehow opalescent like a circle of mother-of-pearl, her light striping the dark, velvety sea. Somewhere, two Scops owls were chiming at each other, like tiny bells in the trees, and the warm heavy air was redolent of sea and flowers and trees.